ACKNOWLEDGEMENTS

The initial collation of information and drafting of this report was undertaken, under contract, by Sir William Halcrow and Partners Ltd. Particular thanks are due to Halcrow's for their contribution to the project.

Subsequent drafting and amendment was undertaken within the Water Management and Science Directorate, NRA Head Office, with substantial contributions and comment from other Head Office Directorates, and from practitioners in NRA Regions. The helpful information, comments and support provided by colleagues are most gratefully acknowledged.

The final drafts benefited from critical appraisal by the Department of the Environment, and their helpful contribution and comments are also gratefully acknowledged.

NRA

National Rivers Authority

DISCHARGE CONSENTS AND COMPLIANCE

THE NRA'S APPROACH TO CONTROL OF DISCHARGES TO WATER

Report of the
National Rivers Authority

March 1994

Water Quality Series No. 17

CONTENTS

LIST OF FIGURES

LIST OF TABLES

EXECUTIVE SUMMARY

There has been considerable public interest in how the NRA undertakes one of its prime functions - discharge control, and in particular, how the NRA interprets its duties and exercises its powers in respect of discharge consents. It is opportune, in the build-up to the proposed Environment Agency, that the NRA should explain how it addresses this vitally important component of water quality management. This is set out in the report.

The report is intended to provide an introduction for the non specialist, covering the evolving legal framework within which the NRA exercises its discharge control powers, the consent setting process, and the results of consent monitoring and enforcement.

Effective control of discharges to water, through the issue, monitoring and enforcement of consents, is fundamental to the NRA achieving its mission of protecting and improving the water environment in England and Wales.

The law controlling pollution of water makes it an offence to cause pollution or permit it to occur. It is also, in most cases, an offence to discharge sewage or trade effluent, without the NRA's consent; and the discharger must comply with any conditions attached to the consent.

When an application for consent to discharge is received, the NRA has to decide whether to grant the consent and, if so, what conditions (if any) should be applied in the consent to protect receiving waters.

Quality objectives, relating to environmental need and end uses of a receiving water, are a fundamental cornerstone of discharge control policy. Once the desired receiving water quality is identified, the necessary effluent quality to ensure achievement of the water quality objective can be determined.

Some current water quality objectives are statutory; others are formal but non statutory. The NRA is in a state of transition as it moves from non-statutory river and estuary quality objectives, towards full development and introduction of Statutory Quality Objectives, and the anticipated introduction in 1994 of the General Quality Assessment scheme for comparative assessment of water quality. As with the current, non statutory river quality objectives, discharge consents will be the major mechanism for achievement of statutory quality objectives.

Discharge consents only apply to point source discharges, that is to say, specific, identifiable discharges of effluent from a known location. Diffuse sources of pollution, such as agricultural run-off, and pollution incidents, such as accidental spillages, cannot be controlled by discharge consents.

The location of the discharge, nature and load of substances discharged, and the dilution available in the receiving water are all important factors in determining the acceptability of a proposed discharge.

Effluents can be broadly categorised as sewage or industrial, and can be sub-categorised according to the nature of potentially polluting substances involved and scale of process.

The Water Resources Act is the principal Act under which the NRA operates, and exercises its discharge control powers. The NRA is in the process of transferring control for discharges subject to Integrated Pollution Control under the Environmental Protection Act 1990, to Her Majesty's Inspectorate of Pollution. EC Directives and International Conventions also have a substantial impact on how discharges are controlled.

The procedures and steps in the consent setting process are briefly outlined and include advertising, consultation, technical consideration, the decision process, appeals, register provisions and review requirements.

The diverse nature and scale of discharges require different types of conditions, which fall into three main categories - numeric, non-numeric and descriptive.

Numeric consent conditions are used for those discharges which have the most potential to exert an undesirable environmental impact, and limits will relate to individual substances or attributes of substances discharged, generally specified as concentrations and/or load, together with flow. Trade effluents are normally subject to absolute limits, whereas municipal sewage effluents are allowed to exceed certain limits in a proportion of samples, subject in some cases to a higher maximum limit.

Water service company discharges account for approximately one third of all consents, mostly sewage discharges. Septic tank discharges account for another third of consents on the Register.

The Urban Waste Water Treatment Directive will, when implemented, make substantial changes to the way in which consents for sewage works are framed and performance monitored. Much debate has centred on achieving UWWTD requirements without incurring unnecessary costs.

The NRA monitors consented discharges and receiving waters in order to assess compliance of the discharger with the consent. If a discharger is found to contravene the terms of his consent, he may be prosecuted.

Legal enforcement is a powerful tool which the NRA uses when necessary to achieve its objectives or when punitive sanction is required for the public good. The NRA is bound by strict rules of evidence, and knowledge of non-compliance with consent does not necessarily result in prosecution. Depending on the reason for non-compliance the NRA will normally exhaust persuasive measures before initiating prosecution.

Given the diversity of discharges and the changes that have been made over time to the regulatory framework, the overall assessment of compliance of a discharger with the terms of a consent can be problematic. The report outlines current compliance assessment methodology, which is likely to continue to develop over the next few years.

Between 1990 and 1992 sewage works compliance in England and Wales rose from 90% to 95%. For industrial discharges compliance was fairly static, with approximately two thirds complying nationally over the same period.

The costs associated with determination of consent applications and of monitoring and assessing compliance with consents are recovered through the Charging for Discharges Scheme.

1 INTRODUCTION

1.1 OBJECTIVES

The NRA's objective in producing this report is to provide information for dischargers and the general public on its approach to the control of discharges to water. In the build up to the proposed Environment Agency, it is opportune that the NRA should, at this time, make more widely known the principles within which it works, and the factors affecting discharge control.

The report is not a legal document, and whilst every effort has been made to ensure accuracy, it must not be taken as being definitive or authoritative with respect to water pollution law. In many places it paraphrases or summarises the legislation, and the original documents should be consulted for precise definitions.

The report outlines:

● the principles of discharge control policy;

● the legal framework and the NRA's operating practice;

● developments in discharge consenting; and,

● discharge, consent compliance, and enforcement statistics for the years 1990 - 1992.

1.2 THE ROLE OF THE NRA

The NRA's mission is to protect and improve the water environment in England and Wales. In undertaking this mission the NRA aims to achieve a continuing overall improvement in the quality of rivers, estuaries, coastal waters and groundwaters, through the control of pollution. An additional aim is to ensure that dischargers pay the costs of the consequences of their discharges, and, as far as possible, to recover the costs of water environment improvement from those who benefit.

Water is used in a variety of domestic, industrial and agricultural processes, and has to be returned to the environment, mostly as effluent discharges. The NRA's role is to ensure that, within its regulatory framework, the right environmental and discharge standards are set, and enforced. Discharge of effluent to a receiving water is acceptable, provided the quality of that receiving water consistently meets or exceeds the minimum requirements set by the NRA or by government.

The regulation of discharges, through the issue, monitoring, review and enforcement of consents, is one of the principal ways in which the NRA fulfils its statutory duty.

The law controlling pollution of water makes it an offence to cause pollution or permit it to occur. It is also, in most cases, an offence to discharge sewage or trade effluent, without the NRA's consent; and the discharger must comply with any conditions attached to the consent.

When an application for consent to discharge is received, the NRA has to decide whether to grant the consent and, if so, what conditions (if any) should be applied in the consent to protect receiving waters.

1.3 GENERAL APPROACH TO CONTROL OF DISCHARGES

The general approach to the control of discharges to surface and groundwaters in England and Wales has been to set discharge limits which are individually based on local circumstances. The alternative is to set uniform emission standards applicable to all discharges of a particular size or type, but this approach, runs the risk of under-protecting the environment or over-specifying the level of treatment necessary for environmental protection. The discharge requirements are therefore tailored to the needs of the receiving water and to key end uses.

Quality objectives, relating to environmental need and end uses of a receiving water, whether statutory or non-statutory, are a fundamental cornerstone of discharge control policy. Once the desired receiving water quality is identified, it is a relatively straightforward process to determine what effluent quality is necessary to ensure achievement of the water quality objective.

1.4   WATER QUALITY OBJECTIVES

Since the late 1970s formally agreed, but non-statutory, river quality objectives have been applied to rivers, largely based on a classification system developed by the former National Water Council (NWC) in which, for example, Class 1a represents good water quality and Class 4, bad. This classification system applies to some 40,000 km of rivers. Other formal, but non statutory, water quality objectives apply to estuaries. The NWC Classification system was used by the NRA's predecessors to identify current water quality and to set targets for improvements so that strategic planning of development could take place, within the overall context of improving or maintaining water quality by using discharge consents set at levels appropriate to achieve the desired river or estuary class. The NRA has continued to use this system when reviewing current water quality and planning improvement.

One of the problems with achievement of NWC river quality objectives, is that they are non-statutory and therefore open to challenge in cases where the NRA's long term quality objectives differ from those of developers or water users. Under the Water Resources Act 1991 the Secretary of State has the power, through Regulations, to introduce a new **statutory** classification system, or systems, and to set water quality objectives based on the classification system, ie target quality classes and dates for their achievement. Having established the classification the Secretary of State may then serve a notice on the NRA, establishing water quality objectives for individual stretches of water. These statutory water quality objectives amount to formal national policy for water quality, which both the Government and NRA are under a duty to achieve.

Some Statutory Water Quality Objectives have already been set, notably those relating to E.C. Water Quality Directives such as the EC Directive on the Discharge of Dangerous Substances to the Aquatic Environment (76/464/EEC). The NRA is required, as directed by the Secretary of State, in these cases, to set conditions in consents to limit any discharges so that the appropriate classification standard is met.

Further use-related water quality classifications, to be applied to defined stretches of water, are under development. The NRA's initial proposals for development of statutory use-related water quality standards, and a general classification system for surface and groundwaters, were outlined in "Proposals for Statutory Water Quality Objectives", (NRA 1991). Following consultation the NRA modified its proposals to Government and aspects of the scheme relating to rivers were published in "River Quality: The Governments Proposals",(DoE, December 1992). Subsequently, in October 1993, the Government issued, for consultation, the draft Surface Water (Fisheries Ecosystem)(Classification) Regulations. The NRA is thus currently experiencing a state of transition as it moves towards full development and introduction of Statutory Quality Objectives, and the anticipated introduction in 1994 of the General Quality Assessment scheme for comparative assessment of water quality. As with the current, non statutory river quality objectives, discharge consents will be the major mechanism for statutory quality objective implementation.

The NRA has also published its Groundwater Protection Policy which aims to ensure that all risks to groundwater resources, particularly point source and diffuse risks to potable supply aquifers, are dealt with in a common framework, applying principles of prevention rather than control where ever possible, "Policy and Practice for the Protection of Groundwater", NRA 1992.

1.5   CONSENTS, COMPLIANCE ASSESSMENT AND ENFORCEMENT

A consent to discharge is a legal document, issued by the NRA, or its predecessors, permitting the discharge of effluent, generally subject to volume and quality constraints, into controlled waters.

Discharge consents only apply to point source discharges, that is to say, specific, identifiable discharges of effluent from a known location. Diffuse sources of pollution, such as agricultural run-off, do not have a specific point of origin or a single person or organisation responsible for the discharge. Diffuse pollution cannot be controlled, therefore, by discharge consents.

Any person wishing to discharge effluent must make an application to the NRA. The NRA will then determine the application taking account of certain criteria, deciding whether consent should be refused or granted either unconditionally or subject to conditions limiting where, when and what may be discharged and specifying monitoring and record keeping requirements. These criteria include the application of NRA policies, and the water quality objectives and associated quality standards for the waters into which it is proposed to make the discharge. In this way the NRA is able to protect the water environment by either refusing to issue consents for discharges which would have an unacceptable impact, or by making consents subject to conditions which will ensure that the quality of the receiving waters is maintained.

The NRA also has discretionary powers to control certain categories of minor discharges such as septic tank discharges to soakaways, which in many cases do not need continuing control. These powers are implemented through issue of Prohibition Notices, which may either be conditional, requiring action by the discharger, or may be absolute, requiring that the discharge stop or not commence.

A consent will only help to protect water quality if the effluent discharged complies with the terms of the consent. The NRA has powers to monitor effluents and receiving water quality, in order that compliance with the conditions of the consent can be assessed. It may also make the provision of discharger's data regarding the quality and volume of the effluent a condition of a consent.

If a discharger is found to contravene the terms of his consent, he may be prosecuted. The NRA will prosecute in appropriate cases, but such enforcement is but one of a number of different methods for achieving the NRA's environmental goals. Further details on consent compliance assessment and enforcement can be found in Sections 9 & 12 respectively.

Consents granted under earlier legislation remain in force even though the Acts under which they were issued have been superseded. Such consents, issued by the predecessor bodies to the NRA, reflect the requirements and limitations of the legislation applicable at that time, and the structure of the consent, or conditions set, may not be appropriate for effective enforcement under the Water Resources Act.

It is a major concern of the NRA to ensure that the consents that it inherited in 1989 are reviewed and revised, taking into account the requirements of the Water Resources Act 1991. This is a substantial task, bearing in mind the total number of consented discharges, (approximately 110,000 in England and Wales), the diversity of discharge and receiving water types, and the different consent register and data recording systems, inherited from the former Regional Water Authorities and currently in use within the NRA regions. The technical and administrative workload associated with such a review has to be balanced against the continuing need to deal with current water quality issues and problems, and to issue and enforce new discharge consents in a climate of changing environmental regulation.

2   TYPES OF DISCHARGE

2.1   NATURE, COMPOSITION, AND ENVIRONMENTAL SIGNIFICANCE OF DISCHARGES

2.1.1   Introduction

Discharges to controlled waters (see Glossary for definition) essentially fall into three distinct categories as follows:

- isolated pollution incidents (eg an accidental spillage of pollutant)
  (Discussed briefly in section 2.1.2. below)

- diffuse pollution (eg agricultural run-off) (Section 2.1.3.)

- point sources discharges (eg from a fixed discharge pipe) (Section 2.1.4.)

Pollution is not defined in UK law and ultimately, it is for the courts to decide on whether a discharge has a polluting effect. A working definition of pollution as used in some European Commission documents is:

"the discharge by man of substances or energy into the aquatic environment, the results of which are such to cause hazards to human health, harm to living resources and to the aquatic ecosystem, damage to amenities or interference with other legitimate uses of water".

Throughout this report the term 'pollutant' will only be used in the context of substances when they exceed the capacity of the receiving water to degrade, dilute or disperse them without harm to aquatic life or legitimate uses made of that water. This is an important point.

2.1.2.   Pollution Incidents

Pollution incidents are often the cause of acute contamination of water bodies, and sometimes can have long term effects. Incidents may occur as a result of accidents, misuse of equipment, or negligence at industrial, commercial, or private domestic premises. They may result from, for instance, the failure of an effluent treatment process, or the unplanned entry of substances into a controlled water either via a surface water drain or directly onto land.

Such incidents cannot be fully controlled by the consenting process, and so fall outside the scope of this report. Control is achieved instead by implementing pollution prevention procedures, or constructing physical structures, on site, and by education of people who handle potentially polluting materials, so as to minimise the risk of incidents occurring, either as a result of process activities or as a result of vandalism. In determining a consent application the NRA takes every opportunity to persuade the discharger of the need for preventative action to minimise the risk of an incident. Consents are structured so that, if complied with, the risk of a pollution incident attributable to routine effluent management activities is negligible. The NRA reports annually on Pollution Incidents to controlled waters (Water Pollution Incidents in England and Wales, NRA, 1990, 1991, 1992).

2.1.3   Diffuse Pollution

Diffuse pollution of surface and underground waters is caused by the entry of pollutants which do not occur through a clearly defined source. Examples of polluting substances entering waters in this way would be:

- contaminated run-off from land (eg pesticides, fertilisers);

- percolation of material to underground and surface waters from contaminated land, landfill sites etc.;

- contaminated rainfall, or deposition or absorption of pollutants from the atmosphere, (eg acid rain)

In the case of contaminated run-off from land it is important to distinguish between true diffuse run-off and that which is purposely drained from land. An airport, for example, would have both diffuse and point source inputs to watercourses. Any pollutants falling on runways and taxiways would be drained via surface water sewers to a watercourse where the output would be through a pipe (a point source) and thus could be treated and consented. Pollutants falling on grassed areas within the airport, would tend to percolate through and across the land to surface waters and/or aquifers and could not be consented. Pollution control would be aimed at preventing discharge of contaminants to these areas.

Diffuse pollution cannot be controlled with discharge consents.

2.1.4   Point Source Discharges

Point source discharges, as the name implies, are those which are discharged through an identifiable source - a pipe, for example. In this case, the concentration and load of pollutants discharged may be measured and the discharge may be amenable to both consenting and treatment measures. Not all point source discharges are controllable. Discharges from abandoned mines for instance, may be polluting yet are outside the scope of the NRA's current consent powers. Highway drains **should** contain only clean surface run-off from roads, yet may convey spillages from road accidents direct to receiving waters.

The consent process is intended to ensure that substances legitimately discharged to controlled waters do not cause pollution.

Pollution incidents from point sources cannot be entirely ruled out, because of the impossibility of planning for every possibility, however unlikely.

Sewage effluent is by far the most numerous type of point source discharge. Industrial effluent discharges fall into 3 main categories, heavy industry, (including power generation, chemicals and petrochemicals,) light industry, (including product manufacture, light engineering and textile manufacture), and the food industry, (including agriculture, beverages and canneries). In addition contaminated site drainage is a special problem associated with surface water discharges

Point source discharges also vary widely in composition and volume, and as a consequence have an equally wide potential effect on the water bodies to which they are discharged. When considering this potential effect the following factors are important.

i   Load of substance or substances in the discharge

The quantity of a substance discharged over a given period is critical in determining the potential of a receiving watercourse to accommodate it without pollution occurring, particularly so for substances which bioaccumulate.

The quantity discharged is normally expressed as a load. This is the product of the concentration of a particular substance multiplied by the volume discharged and is therefore expressed as weight discharged per unit time (eg kilogrammes per day-kg.d^{-1}). Once the load of a substance discharged is known, it is a relatively simple matter, (as briefly discussed in section 4.2.3) to calculate the resulting increase above background concentration in a stream by dividing the load discharged per unit time by the river flow during the same period.

ii   Dilution available in the receiving water body

The dilution available in the receiving watercourse is an important factor which affects the concentration of a substance in the aquatic environment. When a discharge is to a watercourse where the flow is uni-directional (eg a river) calculation of the concentration of substances downstream of the discharge is relatively straightforward. It is more difficult to assess where the flow of the receiving water body is multi-directional, (eg canals, estuaries and coastal waters,) or unknown, (eg underground waters).

The impact of a substance can be felt at a number of levels, for example as a concentration, which may have an acute impact on aquatic life over a short period of time, or as a load, giving chronic accumulative impact over a long time period. In both cases the potential for dilution to harmless concentrations is an important factor in considering whether a point source discharge should be permitted.

iii   Position of the discharge

The position of the discharge, both geographically and in relation to other discharges, is important as it will indirectly affect the amount of dilution water available for other discharges in the receiving watercourse.

iv   Nature of the polluting substance

The nature of a potentially polluting substance and thus its effect on the aquatic environment will have a major influence on the amount which will be allowable in any particular water body. When considering this effect a number of important aspects of the substance must be assessed including:

● acute and chronic aquatic toxicity

● toxicity to higher organisms (including man)

● bio-accumulation potential

● persistence in the aquatic environment

● biodegradability

● solubility

● visibility

All of the above, and other considerations, in particular the uses made of the receiving waters, will determine the acceptability of the substance in the watercourse, based on its effect on the aquatic environment and its ultimate fate. This acceptability is termed a water quality standard, and is usually defined in terms of a concentration of the substance in the receiving water at which no harm will occur for an identified use such as ecosystem protection or abstraction for potable supply. The development of appropriate water quality standards is a complex task. Existing water quality standards mainly originate from EC legislation and the existing NWC river and estuary quality classification scheme.

The NRA is progressing with the necessary research to develop a range of appropriate water quality standards to be included in possible classification schemes for use in related water quality objectives for controlled waters.

As effluents continue to become more complicated in composition, it becomes increasingly more difficult to establish appropriate water quality standards, particularly where individual components may react together or interact to exert a more toxic effect. The NRA is also undertaking research to investigate the feasibility of toxicity-based consents, which would use a direct measure of toxicity of the whole effluent to control discharge impact, rather than rely on the specification in consents of discharge concentration limits for a large suite of individual substances. If the research is successful, and such consents are introduced, there would still remain a requirement for individual limits for certain dangerous substances in order to comply with EC requirements.

2.2 SEWAGE

2.2.1 Introduction

Discharges of sewage may arise from a number of different sources including:

- treated effluent and storm discharges from urban sewage treatment works

- effluent from sewage treatment plants serving small populations, known as "package plants"

- combined sewer overflows and emergency overflows from sewerage systems

- septic tanks

- crude sewage discharges at some estuarine and coastal locations

2.2.2 Sewage Treatment Works

Most of the sewage effluent entering surface waters in England and Wales is discharged from sewage treatment works (STWs) run by the ten Water Service Public Limited Companies (WSPLCs). The quality of sewage effluent from such works varies according to the treatment processes used, which can normally be divided into the following general categories:

- preliminary treatment (removal of rags, grit, gross solids etc.)

- primary treatment (settlement of solids in suspension)

- secondary treatment (biological treatment to remove organic material)

- tertiary treatment (to decrease any remaining solids or organic material or to provide disinfection or nutrient removal)

Normally, sewage treatment plants are designed to perform the treatment of sewage in the order set out above, although few plants in England and Wales need to proceed beyond the secondary treatment stage. Some sewage works, mainly those in coastal or estuarine locations, only perform primary treatment on sewage received, or discharge preliminary treated effluent direct to sea via a long (offshore) outfall where natural forces are used to dilute and disperse the effluent.

2.2.3 Package Sewage Treatment Plants

These are, in general, much smaller units than urban sewage treatment works, providing secondary treatment within a self-contained plant. They are used for small domestic housing developments and commercial or industrial premises where connection to a main sewer is not practical or economic. Providing they are operated correctly and serviced at the recommended intervals, most are capable of producing adequate quality effluents.

2.2.4 Intermittent Discharges - Emergency, Combined Sewer and Storm Tank Overflows

In order to prevent the flooding of properties by sewage, urban sewers are designed with overflows, whereby sewage which cannot be passed through the sewerage system is discharged direct to a watercourse. These discharges, although mostly infrequent, can result in significant volumes of sewage being discharged to receiving watercourses. NRA controls over such intermittent discharges are intended to minimise their impact, and to ensure that they only occur when necessary.

i   Emergency Overflows

These discharges are normally made from sewage pumping stations following a pump or power failure. To prevent sewage flooding the surrounding area, an overflow pipe carries sewage to a nearby watercourse. The discharge from such an installation consists of crude sewage, and is only legal in defined emergency conditions. In such cases, it is recognised that the possible harm to the receiving water as a result of the discharge is outweighed by the potential risk to health and property that would be caused by foul sewage flooding.

ii   Combined Sewer Overflows

Most urban sewerage systems are designed to carry rain water draining from roads and roofs in addition to sewage ('combined sewers'). However, it is impractical to install sewerage systems which will cope with all rainfall intensities, and they become hydraulically overloaded during more extreme rainfall events. To prevent flooding of properties and hydraulic overload of the STW the weak, excess, 'storm sewage' is allowed to overflow to a nearby watercourse where the dilute sewage is expected to be accommodated, without pollution occurring, by the increased flow in the receiving water due to the same rainfall event. The quality, and impact, of such discharges depends upon the rainfall intensity and the design of the overflow, especially the setting of the overflow weir, and the efficiency of the system in retaining aesthetically objectionable solids.

iii   Storm Tank Overflows

In wet weather flows to sewage treatment works increase, particularly if the works is served by a combined sewer. Usually a works is designed to treat a certain multiple of the flow it receives in dry weather, typically three times dry weather flow. Above this flow the works may not be capable of treating greater volumes of sewage adequately, and so excess flow is diverted at the sewage works inlet to storm tanks designed to retain such flows for subsequent treatment at the sewage works once the flow in the sewer has returned to normal. Should the wet weather persist, the storm tanks will eventually fill and begin to overflow. To prevent flooding, this storm tank discharge will be directed to a nearby watercourse. The storm tank discharge will consist of dilute sewage, after some settlement of suspended, potentially polluting, material.

2.2.5   Septic Tanks

Septic tanks are small sewage treatment facilities which normally serve individual domestic premises. They provide primary settlement of sewage together with some anaerobic digestion of the collected sludge. Effluent from septic tanks can be of reasonable quality if the appropriate maintenance has been carried out. In England and Wales most septic tank effluents are discharged to underground strata by means of a soakaway.

2.2.6   Crude Sewage Discharges

Some sewage outfalls, mostly in coastal or estuarine areas, discharge close to the shore, and are not subject to treatment of any kind, or receive only preliminary treatment consisting of some form of screening. Other, more recently constructed outfalls, are much longer, often extending several kilometres off-shore, to make use of natural dilution and dispersion for preliminary treated effluent.

Such discharges consist of crude (or screened or macerated) domestic sewage and sometimes include trade effluents. On implementation of the Urban Waste Water Treatment Directive, (as further described in Section 7,) many of these discharges will cease, with the sewage flow being directed to new sewage treatment works. Some existing outfalls will be retained as discharge points for treated sewage, or as storm or emergency overflow discharges.

2.2.7   Potentially Polluting Substances Discharged

The substances discharged in sewage effluent may include the following:

- solid material which can coat the bottom of rivers, smothering aquatic life, fish breeding grounds etc. (as measured by the suspended solids (SS) content of the discharge)

- oxygen-demanding material which removes life-supporting oxygen from the receiving watercourse (eg organic material; the likely effect of which can be measured by the biochemical oxygen demand (BOD), or chemical oxygen demand (COD), of the discharge)

and may include limited amounts of:

- toxic material which can poison aquatic life (eg ammonia, heavy metals, toxic inorganic or organic compounds)

- potential human or animal pathogens (eg certain bacteria or viruses which are unaffected by the treatment process).

For some sewage works much of this material originates from trade effluents discharged to sewers. In applying for a consent for the STW discharge the sewage works operator is obliged to disclose the likely presence of any toxic components in the sewage. The NRA discharge consent may specify limits for such substances, and the sewage works operator has to ensure that any trade effluent to sewer consents, or agreements made to discharge to sewer, do not permit exceedance of STW effluent limits, and that dischargers to sewer comply with the terms of their trade effluent consent.

2.3   HEAVY INDUSTRY

2.3.1   Introduction

The term 'heavy industry', although not a rigorous definition, is generally applied to large, manufacturing, processing or extraction plants. Such industries are perceived as having a high potential for causing environmental pollution and are targeted, together with certain elements of light industry, for rigorous process control by Her Majesty's Inspectorate of Pollution, through implementation of Integrated Pollution Control (IPC), under the terms of the Environmental Protection Act 1990. The requirements of this Act are briefly summarised in Section 3.2.4. Examples of the industrial sectors involved are coal and bulk mineral extraction, power generation, steel manufacture, chemical industry, and petrochemical refining and production.

2.3.2 Potentially Polluting Substances Discharged

Coal and bulk mineral extraction

● **Inert solids** from spoil tips and roadways.

● **Acidic minewater** pumped from the mine to allow extraction of minerals.

● **Washery effluent** can contain phenolic oil which can cause a high oxygen demand and lowering of pH.

Power Generation

● **Excess heat** in cooling water discharged can cause thermal shock to fish, and also reduces the oxygen carrying capacity of water.

● **Antifouling chemicals** used to prevent slime build up, etc. Excess dosing may result in the discharged cooling water retaining a toxic effect.

● **Stock yard drainage** - chiefly acid run off from stocked coal and suspended solids.

Steel Manufacture

● **Stock yard drainage** contains mainly suspended solids. pH and sulphides may be a problem with some blast furnace slag leachates.

● **Gas plant effluent** has a high oxygen demand and ammonia, and cyanide content, together with phenolic tar residues and complex organic chemicals.

● **Blast furnace gas scrubbers** remove metalliferous dust and combustion gases by wet scrubbing and the resultant effluent requires settlement and pH adjustment.

● **Initial Steel Processing** is a chiefly 'dry' process, but spillage or weepage of oils and chemicals can cause substantial oxygen demand, and toxic effects.

● **Further processing** involving acid pickling and washing processes can result in a large volume of acidic effluent, requiring neutralisation of pH, and removal of metal content and suspended solids. Weepage or spillages of lubricants or process chemicals can also occur.

Chemical Industry

The chemical industry is extremely diverse, in scope and scale, and it is outside the scope of this report to address such a complex issue. The 'heavy industry' component generally comprises the large scale bulk manufacture of chemicals which may be used in their own right, or as components of further, smaller scale, refined products. Each bulk chemical processing or manufacturing business, such as potable water treatment, or fertiliser, paint, and pharmaceutical manufacture, has its own specific features. Commonly pH, suspended solids, metals and toxic by-products need to be controlled.

Petrochemical Refining and Production

There is a large range of processes, and again, each has its specific features. Common problems affecting effluent quality include oxygen demand of process streams, pH control, immiscible and soluble oil content of effluent, toxic by-products, and containment of spillages on site of raw material or product.

2.4 LIGHT INDUSTRY

2.4.1 Introduction

The term 'light industry', although again not a rigorous definition, is normally applied to industrial processes which are carried out on a smaller scale than those outlined in Section 2.3.1. Depending on the nature of the processes involved individual plants may come within the scope of IPC. A range of businesses to which this definition applies would include light engineering and fabrication, product manufacture, tanneries, paper mills, metal finishing, textile manufacture, and mineral processing.

2.4.2 Potentially Polluting Substances Discharged

The range of potentially polluting substances which could be discharged is enormous, and would include those given in Section 2.3.2, although generally in smaller quantities. Other contaminants which often prove problematic include detergents, dyestuffs, biocides, and metal finishing liquors.

2.5 FOOD INDUSTRY

2.5.1 Introduction

The food industry encompasses both primary (agricultural) production and the secondary processing of food.

2.5.2 Agriculture

Only certain elements of agricultural production lead to point source discharges which may be amenable to treatment and control via the consents mechanism. In appropriate locations this could include, for instance, milking parlour washings, and effluent from intensive meat production units, but normally effluents from livestock production cannot be economically treated to a standard suitable for direct discharge to water. Such wastes have a very high organic content and oxygen demand, but are useful fertilisers and are best disposed of through careful application to agricultural land, in order to avoid river pollution, and in accordance with the Code of Good Agricultural Practice issued by the Ministry of Agriculture and Welsh Office.

Certain food production techniques, such as fish farms and cress farms, rely on the aquatic environment itself in order to produce marketable products. They rely on high quality abstracted water, but the intensive production processes may lead to a deterioration in quality of the water returned to the river or stream, and discharges are therefore subject to control through consents.

Arable farming leads to few point source discharges which are economically amenable to treatment. Large farms may use treatment plant for removal of pesticide residues from application equipment. Such plant normally discharge effluent onto land and are controlled through issue of prohibition notices. On most arable farms dilute pesticide washings can be safely disposed of onto land in accordance with the Code of Good Agricultural Practice.

Other agricultural activities may lead to diffuse inputs, such as surface run-off following application of fertilisers or pesticides, or to pollution incidents, such as leaking silage liquor storage tanks or slurry lagoons.

2.5.3 Food Processing

The food processing industry encompasses a very wide range of processes and factory units, including slaughterhouses, creameries, breweries, drinks manufacturers, fresh and frozen vegetable preparation, canning factories, sugar refiners and fish processors.

Despite the breadth of merchandise and processes represented by the industry, effluents tend to contain a large quantity, or 'load' of readily biodegradable organic substances. These substances are capable of decreasing the amount of dissolved oxygen in the receiving water to which they are discharged, and must normally be removed or reduced by treatment prior to discharge. As outlined in Section 2.2.7, the potential demand for oxygen is assessed by measuring the BOD or COD of the discharge.

Other effluents from the food industry may typically contain suspended solid material, cleaning agents, and bactericides and excess heat.

2.6 SITE DRAINAGE

2.6.1 Introduction

As mentioned in Sections 2.1.3 and 2.1.4, drainage from sites used for industrial, commercial or domestic purposes may be collected in a surface water drainage system discharging as a point source to a receiving water, or may enter ground or surface waters by percolation as a diffuse input. Generally, site drainage consists of rainfall which has fallen on the site but which may have become contaminated by passing over or through contaminated material.

For most sites the following types of site drainage discharge are possible:

- to a public or private surface water sewer discharging to a receiving water;

- to a public or private foul sewer discharging to a sewage treatment works;

- direct to a surface watercourse via a pipe, culvert or drain;

- to underground strata, and therefore groundwaters, via direct percolation, a soakaway or a land drainage system.

2.6.2 Potentially Polluting Substances Discharged

Potentially polluting substances discharged from the site drainage systems of a particular industry will reflect the chemicals and materials used by that industry in its business. Fuel (particularly oil products), raw and finished materials, and cleaning chemicals are all typical contaminants of such discharges.

Where these materials are discharged through a defined outfall, it is possible to control their concentrations by means of a discharge consent. Conditions in such a consent would depend on the contaminants likely to be discharged.

If considered necessary, based on the perceived risk, some treatment may be installed to prevent unwanted substances reaching the aquatic environment. Most commonly, treatment for such discharges consists of the installation of an oil interceptor to remove oil or petrol, or a settlement tank to remove suspended material.

3   LEGAL BASIS FOR CONTROL

3.1   HISTORY OF WATER POLLUTION LAW

3.1.1   Background

Water pollution legislation has developed steadily over the last century but has origins dating from 1388, when water pollution was prohibited by the Act for Punishing Nuisances which Cause Corruption of the Air near Cities and Great Towns. This, along with other early legislation, had, as its main aim, limiting the generation of foul smells from putrefied waters. The link between water pollution and outbreaks of disease had yet to be made.

Increasing urban populations and lack of engineering skills compounded the problems of pollution by sewage in cities and towns. During Elizabethan times the stinking streets of London, caused by a mixture of unsanitary practices including sewage disposal, were a characteristic of that era. To combat the problem, pomanders and other scents were used prolifically, and women wore pattens, (wooden clogs or sandals with a raised wooden platform,) to raise them above the filth accumulating in the streets. Diseases such as bubonic plague, dysentery, typhus, and typhoid were endemic.

The rapid growth of industrialisation in the early 19th century prompted further legislation on industrial water pollution. This related to specific industrial activities which were recognised as being the most significant causes of pollution, such as, for instance, gasworks,(which produce particularly noxious effluent), or harbours, docks, and piers, (by controlling the emptying into water of ballast or other substances which cause pollution or hazard to shipping.)

With advancing engineering techniques the early 1800s saw the construction of sewerage systems in many of the larger towns. These were originally intended for draining surface waters and it was forbidden for domestic wastes to be discharged into them; cesspits were the main means of disposal of domestic waste. Discharges from cesspits constituted a major problem causing extensive water pollution and in 1847 a law was passed requiring discharges from cesspits to be put into the foul sewers.

The consequence of this legislation was far reaching, particularly in London where all the sewers discharged directly into the Thames - which at that time was a major source of water for the city. Furthermore the poor state of repair of the sewers themselves resulted in contamination of the ground water, London's second main source of water. Cholera and typhoid epidemics ensued and it was finally established, following the Broad Street Pump cholera outbreak in 1854 in which 10,000 deaths were reported, that these were directly related to polluted drinking water. Water borne diseases in London were finally brought under control in 1870, with the completion of Bazalgette's new sewerage system and the abstraction of water for potable use from Teddington, upstream of the polluted reaches.

3.1.2   Relevant Anti-pollution Legislation

The main Acts of Parliament, over the past century and a half, which have been intended to ensure effective control of discharges to the water environment, are listed below.

Public Health Act 1848
Salmon Fisheries Act 1861
Rivers Pollution Prevention Act 1876
Public Health Act 1936
Rivers Board Act 1948
Rivers (Prevention of Pollution) Act 1951
Clean Rivers (Estuaries and Tidal Waters) Act 1960
Rivers (Prevention of Pollution) Act 1961

The Water Resources Act 1963
The Water Act 1973
The Control of Pollution Act 1974
The Water Act 1989
The Environmental Protection Act 1990
The Water Resources Act 1991

Brief details of these Acts are given in Appendix A.

3.1.3   The Change in Approach to Water Pollution Control

Much of the early legislation on water pollution was created to address specific problems of nuisance or injury as they arose, such as discharges from industrial activities, cesspits etc. As a result a rather piecemeal approach to water pollution control developed. The link between polluted potable water supplies and public health, made in the 1840s and 1850s, was a major factor in subsequent legislative developments, with improvement in environmental water quality being seen as a welcome, but not essential consequence.

In 1912 the Royal Commission on Sewage Disposal published its seventh report, which established the principles of establishing the river dilution available for an effluent, and advocated the general adoption of emission standards applicable to certain ranges of dilution of sewage effluent. The so called Royal Commission standard of 20 mg.l^{-1} BOD and 30 mg.l^{-1} suspended solids, for river dilutions greater than eight times, became the norm for secondary treatment works, albeit non statutory. The 1912 report can be seen as a significant step towards the NRA's current approach to water quality and discharge control.

With increased public aspiration for a cleaner environment, the main trend, evident in the development of the more recent legislation, is that the Acts have no longer addressed one particular problem but have increasingly adopted a more comprehensive approach to pollution control. That is, the legislation is now formulated so that it addresses the source and nature of pollution, the environment to which it is discharged, and the resultant impact upon that environment, the emphasis being upon pollution prevention and control, rather than remediation.

3.2   CURRENT LEGISLATION

3.2.1   Introduction

Current UK legislation for controlling discharges to the water environment has, as outlined in Appendix A, been the result of much debate and modification. Recent legislative changes have consolidated many pollution control statutes within the Water Resources Act, the main elements of which are outlined below or in the sections which follow. Those elements of other relevant legislation which may impact upon discharge control are also addressed below.

3.2.2   Water Resources Act 1991

The Water Resources Act (WRA) provides the statutory basis for all of the NRA's functions and responsibilities. It defines the powers, duties and obligations of the sponsoring Ministers and of the NRA.

The Act deals with the control of pollution of water resources. It introduces powers for the Secretary of State to introduce Statutory Water Classification Systems and to set Statutory Water Quality Objectives. It also establishes offences relating to the pollution of controlled waters and defences against such offences, the chief defence being the possession of, and compliance with, a discharge consent.

Other sections of the Act provide for measures to prevent and control pollution, procedures for consents for discharges, and requirements for samples when used as evidence. Further provisions cover charging schemes, by which the NRA can recover costs incurred in the issue and monitoring of consents, and the maintenance of a Public Register.

These issues are covered elsewhere in this report or in the sections which follow.

3.2.2.1 Pollution Offences (Sections 85 and 86)

The principal pollution control provision in the Act is section 85, which creates the following offences:

i   to cause or knowingly permit "any poisonous, noxious or polluting matter or any solid waste matter" to enter controlled waters.

ii   to cause or knowingly permit any trade effluent or sewage effluent into "controlled waters".

A person found guilty of committing an offence will be liable, following summary conviction, to a fine of up to £20,000 and/or imprisonment for up to three months; or following conviction on indictment, to an unlimited fine and/or imprisonment for up to two years.

There are a number of exceptions to these overall pollution offences. In particular, under Section 88 of the Act, no offence is committed if a discharge is made in accordance with a consent granted by the NRA. The determination of applications for discharge consents is one of the NRA's principal pollution control activities. The onus is, however, on the discharger to obtain the necessary consents. The NRA may give consent with or without conditions, or refuse consent; and also has the power, subsequently, to modify or revoke consents.

The NRA can also issue a Prohibition Notice imposing conditions on the discharger, prohibiting or placing conditions on certain types of discharge such as discharges "onto or into land". Breach of any condition is an offence.

The NRA thus has the discretion to impose controls over supposedly "clean" discharges to watercourses, or over discharges to land (such as those from septic tanks to soakaways), which would not otherwise be subject to control unless pollution resulted.

3.2.2.2 Allowable Defences to Section 85 Offences

Whilst possession of a consent is the main defence, statutory licences and authorisations administered by other regulators may also provide a defence.

Other defences for specific circumstances are available through section 89 including discharge made in the event of a defined emergency, any discharge of trade or sewage effluent from a vessel, or discharge from a highway drain unless a prohibition notice applies.It is not an offence to permit the entry of waters from an abandoned mine to enter controlled waters, nor, in closely defined circumstances, is it an offence to allow mining or quarrying waste to enter controlled waters.

3.2.2.3 Applications for Consent to Discharge (Section 88)

Sections 88 and 109 and Schedule 10 of the Act are concerned with the procedures for consent applications, conditions imposed by the consent, and requirements for a public register to be maintained. Their detailed requirements are addressed in Section 4; Consent Setting Procedure.

3.2.2.4 Supplementary Provisions with Respect to Water Pollution

Within Sections 131 - 132 of the Act, provisions are made for charging for the work undertaken in connection with consents. The scheme of charging is outlined in Section 11.

Other Sections of the Act deal with preventative or mitigating action (Section 161), powers of access (Section 169), and evidence of samples (Section 209). The latter issue is referred to in Section 12; Enforcement for Non-compliance.

Under Section 169 of the Act entry can be made into premises or onto land by any authorised person in order to monitor consent compliance or to investigate pollution. NRA officers are empowered to carry out inspections, measurements and tests and to take away samples of water or effluent.

3.2.3 Rivers (Prevention of Pollution) Acts 1951 and 1961

Although the 1951 Act was, itself, entirely repealed under the Control of Pollution Act 1974, certain byelaws made under the 1951 Act remain in force. These relate specifically to pollution from vessels fitted with sanitary appliances. Such byelaws may allow for the prohibition or regulation of the keeping or use of vessels on specified waters.

The only parts of the 1961 Act still remaining in force are concerned with the sampling of effluent and the establishment of two presumptions regarding the sampling point, and a section relating to restrictions on the disclosure of information. The first presumption relates to the location of the sampling point for admissibility in any legal proceedings. The assumption is that the consent conditions relate to the effluent at the end of the pipe, that is, the entry point to the watercourse. It may not be physically possible to obtain a representative sample at this point and, as a consequence, the provision of a manhole or a sampling chamber at a point in the effluent pipe remote from the point of discharge, will be a condition of the consent. The presumption is that samples taken at this point will be indicative of the final effluent and will be used for assessing consent compliance.

The second presumption relates to consents where no sampling facility is provided. In these circumstances an agreement may be made between the discharger and the responsible authority as to the location of the effluent sampling point. Samples taken at this point may then be used in legal proceedings, and will be presumed to be representative of what was discharged unless shown to the contrary.

3.2.4 Environmental Protection Act 1990

The 1990 Environmental Protection Act introduced a significant change to the way in which controls from particular activities are established. This requires control of processes as well as discharge quality through the application of the principles of Integrated Pollution Control (IPC). This approach to pollution control applies to the discharge of substances from the most potentially polluting processes (prescribed processes) into all environmental media (air, land and water). Control of prescribed processes is divided into two parts; Part A for control by Her Majesty' Inspectorate of Pollution (HMIP) and Part B for Local Authority air pollution control.

The main objectives of IPC are twofold. The first is to prevent or minimise the release of prescribed dangerous substances, (Table 3.1,) and to render harmless any such substances which are released. Other substances released should also be rendered harmless. The second objective is to develop an approach to pollution control that considered discharges from industrial processes to all media in the context of the effect on the environment as a whole.

The Act requires the use of BATNEEC (Best Available Techniques Not Entailing Excessive Cost) to prevent, minimise or render harmless the release of substances into the environment. Where release to more than one environmental media may occur, the Best Practicable Environmental Option (BPEO) must be adopted so that the releases have the least effect on the environment as a whole.

A phased programme for implementation of IPC has been developed (Table 3.2) and, for prescribed processes, HMIP may take over from the NRA responsibility for controlling discharges to receiving waters and assessing discharge compliance.

HMIP is responsible for issuing appropriate conditions for prescribed processes. However, where the prescribed process results in a discharge to controlled waters, the NRA must be consulted, and any relevant requirements of the NRA, including discharge limits and provision of effluent monitoring data, must be inserted into the authorisation. The NRA retains responsibility for monitoring the quality of the receiving waters, and may recover the costs incurred, both in processing Authorisation applications and in undertaking routine environmental monitoring, via HMIP.

In situations where the effluents from a prescribed process and from a non-prescribed process discharge from the same pipe, a consent from the NRA may be required in addition to the authorisation from HMIP.

Table 3.1

Prescribed Substances for Release to Water	
Mercury and its compounds	1,2-Dichloroethane
Cadmium and its compounds	All isomers of Trichlorobenzene
All isomers of hexachlorocyclohexane	Atrazine
All isomers of DDT	Simazine
Pentachlorophenol and its compounds	Tributyltin compounds
Hexachlorobenzene	Triphenyltin compounds
Hexachlorobutadiene	Trifluralin
Aldrin	Fenitrothion
Dieldrin	Azinphos-methyl
Endrin	Malathion
Polychlorinated Biphenyls	Endosulfan
Dichlorvos	

3.3   EC DIRECTIVES AND INTERNATIONAL CONVENTIONS

EC Directives are legally binding on Member States, and through ratification by the Council of Ministers, Member States have committed themselves to achieve the individual Directive requirements. Directives relating to the water environment have a substantial impact on how discharges are controlled, and form a cornerstone of UK discharge control policy. The UK generally implements the environmental Directives through Regulations issued as Statutory Instruments, although some earlier Directives remain implemented through administrative measures. The Regulations identify the competent regulatory authority, and the actions required of it in order to achieve the Directive's requirements.

International conventions do not necessarily have the status of law, but, through signature and ratification, members of the convention have signified their commitment to the convention's goals.

Brief details of EC Directives and International Conventions that have a direct or indirect bearing on discharge control are set out in Appendix B - Summary of EC Directives and International Conventions.

Table 3.2 Timetable for Implementing Integrated Pollution Control.

Sector	Process	Comes Within IPC
Fuel & Power	Combustion (>50 MWH)	1 April 1992
	Boilers & Furnaces	
	Gasification	
	Carbonisation	
	Combustion	
	Petroleum	
Waste Disposal	Incineration	1 August 1992
	Chemical Recovery	
	Waste Derived Fuel	
Minerals	Cement	1 December 1992
	Asbestos	
	Fibre	
	Glass	
	Ceramic	
Chemical	Petrochemical	1 November 1993
	Organics	
	Chemical Pesticide	
	Pharmaceutical	
	Acid Manufacturing	
	Halogen	
	Chemical Fertiliser	
	Bulk Chemical Storage	
	Inorganic Chemical	1 May 1994
Metal	Iron & Steel	1 January 1995
	Smelting	
	Non-ferrous	1 May 1995
Other Industry	Paper Manufacture	1 November 1995
	Di-isocyanate	
	Tar & Bitumen	
	Uranium	
	Coating	
	Coating Manufacturing	
	Timber	
	Animal & Plant Treatment	

4 CONSENT SETTING PROCEDURE

4.1 STEPS

4.1.1 General

Figure 4.1 shows a generalised flow diagram illustrating the consent setting procedure. Although there are some variations in detail between NRA regions in their procedures for processing consent applications, the basic steps, described below, are common.

4.1.2 Preliminary (or Pre-application) Consultations

Often an applicant will have a tentative proposal and approaches the NRA to identify the environmental requirements that may be imposed upon the discharge. The discharge options are reviewed with the discharger prior to submission of the formal consent application, and the NRA tries to provide guidance to the discharger on the likely limiting environmental criteria and any information requirements that the discharger will have to satisfy if the future application is to be determined. However, throughout such discussions the NRA has to ensure that such advice is provided without prejudice to its statutory regulatory responsibilities.

Such preliminary consultation, in advance of the submission of a consent application, is an important component of the consenting process, especially for large and complex discharges. It is often beneficial to the developer to resolve likely discharge requirements, before committing expenditure on detail scheme design.

The NRA uses preliminary consultation as an opportunity for the discharger to explore waste minimisation and pollution prevention measures directly or indirectly associated with the process(es) leading to effluent discharge.

4.1.3 Planning Consultation

Although it is the discharger's responsibility to obtain the necessary consents, the NRA maintains close liaison with local planning departments, in order that developments which might require discharge consent applications are identified at the earliest possible time, and so that pollution prevention measures can be assessed. Local authority planning consultation provides an essential mechanism for the NRA to obtain initial information about proposed developments that may affect water quality or result in an application for discharge, and also, to seek to influence the planning decision, particularly with regard to planning consent conditions for pollution prevention measures. Such planning conditions could, for instance require the provision of bunds around oil tanks, or the provision of oil interceptors on car parks. From the information obtained through the planning process the NRA may contact the developer direct to establish the details of the proposed development, or may comment to the local authority identifying requirements, or both.

The NRA is a statutory consultee on certain planning applications and will be consulted by the relevant local authority for such applications, although the local authority is not bound to follow the NRA's recommendations in determining the application.

4.1.4 Application

Prospective dischargers can obtain a standard application form from the NRA. General guidance notes are available, and NRA officers will provide specific guidance as requested.

Figure 4.1 Consent Setting Procedure

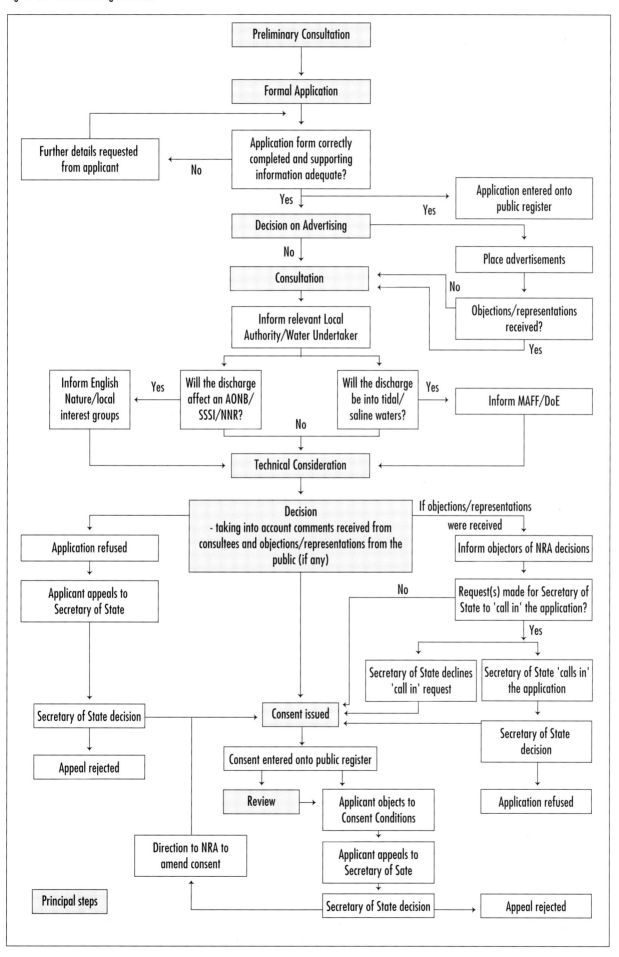

In addition to a completed application form, the NRA can reasonably require the applicant to provide additional supporting information or documentation which, in the case of large or particularly significant discharges (for example, potentially highly polluting effluents or discharges to sensitive areas), can be quite extensive. Such information might include detailed effluent quality and flow data, process and/or waste stream details and monitoring facilities. Additionally, normally only for major schemes or particularly sensitive sites, the NRA may require the applicant to undertake a focused environmental assessment to establish the likely impact of the proposed discharge. The NRA will specify its information requirements. It is for the applicant to undertake or commission the necessary investigations to the quality standards laid down by the NRA.

4.1.5 Advertising

Upon receipt of an application, the first step in the consenting process is for the NRA to consider the likely impact of the proposed discharge on the receiving water so that a decision can be made as to whether public advertising of the application is required. If it is considered that the proposed discharge will have "no appreciable effect" on the receiving water, the NRA may waive the advertising and certain consultation requirements.

The criteria set for establishing "no appreciable effect" for a potential discharge are threefold, and were originally set out in DoE circular 17/84 - The Implementation of Part II of the Control of Pollution Act 1974. The NRA continues to work to such principles, namely that advertising may be waived if:

(a) the discharge does not affect an area of amenity or environmental significance, ie beaches, fisheries, Sites of Special Scientific Importance etc; and

(b) there will be no major change in the flow of the receiving waters; and

(c) there will be no change in water quality such as to impair existing or future uses, or a 10% or greater change in concentrations of any key water quality parameters.

Where the NRA considers that the discharge may have an appreciable effect it must:

(a) publish notice of the application, at least once in each of two successive weeks in an appropriate local newspaper or newspapers; and

(b) publish a copy of the notice in the London Gazette.

The costs of advertising are recovered separately from the applicant.

A person may write to the NRA within six weeks of the notice being published in the London Gazette, making representations or objections. The NRA has a duty to consider all such responses.

The Secretary of State may grant exemption from advertising requirements to the applicant if he is satisfied that it would not be in the public interest to do so or that a trade secret may be jeopardised.

In practice, relatively few consent applications are advertised, as the majority of applications are for small discharges which are considered to exert 'no appreciable effect'.

4.1.6 Consultation

In addition to advertising, the NRA is required to consult various bodies, and to send them copies of the application so that they may comment on the proposed discharge. Statutory consultees are the local authority and water undertaker, within whose area any proposed discharge is to occur; English Nature, (or in Wales, the Countryside Commission for Wales), National Parks and Broads Authority regarding notified sites of special interest; and, for discharges to tidal waters, the Ministry of Agriculture, Fisheries and Food, (or, in Wales, the Welsh Office Agriculture Department). At the request of MAFF and WOAD copies of applications for discharges to tidal waters are also sent to the relevant Sea Fisheries Committee for consideration. The Department of the Environment (or, in Wales, the Welsh Office Environment Department) also receives a copy but does not comment on the merits of the case as this could prejudice the relevant Secretary of State's position in determining any subsequent appeal.

The NRA may also notify other external bodies such as Her Majesty's Inspectorate of Pollution, and local interest groups as appropriate.

If the NRA considers the discharge will "have no appreciable effect" it may disregard the consultation requirements with the exception of copies to be served on Ministers.

4.1.7   Technical Consideration

When an application for consent to discharge is received, the NRA has to consider whether to grant the consent, either unconditionally or subject to conditions, or whether to refuse it. If a consent is to be granted it must be within four months of the date on which the application was received, or such longer time as agreed with the applicant. An application is deemed refused if it has not been granted within the four month period and no extension has been agreed.

The NRA may impose such conditions on the consent as it thinks fit. These may relate to:

● the location of the discharge;

● the design and construction of any outlets;

● the nature, origin, composition, temperature, volume and rate of the discharge;

● periods during which the discharge may occur;

● steps required to minimise the potential polluting effects of the discharge;

● facilities for sampling;

● requirements for in situ monitoring equipment;

● maintenance and provision of records and information.

For most discharges the main criterion against which the NRA considers a consent application is the need to meet the water quality objectives and appropriate standards for the waters into which it is proposed the discharge should be made. Brief details of the technical consideration process are given in Section 4.2.

4.1.8   Decision

Following the technical consideration a decision is reached on whether to grant the consent, and if so, what conditions, if any, will be imposed.

For those applications where the NRA proposes to issue a consent, and which have attracted written representations or objections, the NRA is required, under current legislation, to notify respondents before it may issue the consent. In practice, the NRA, in doing so, generally informs such people of its proposed consent conditions (together with its reasons for issuing the consent). The objectors are informed that they have 21 days within which to request the Secretary of State formally to "call-in" the application, ie to take the decision out of the hands of the NRA, and decide it himself. The NRA cannot decide the application until the 21 day period is up; and if requests to the Secretary of State are made, and notified to the NRA, then the NRA cannot proceed until the Secretary of State has decided whether or not to intervene.

These procedures for the making of call-in requests do not amount to a mechanism by which objectors can formally appeal against the NRA's proposed decision: they simply provide a means for objectors to bring cases to the attention of the Secretary of State, and his responsibility is limited, at that stage, to considering whether or not the application should be called-in. If he decides that a case does not warrant such action, the responsibility for the decision remains with the NRA.

Regardless of this specific provision which allows requests to be formally made to the Secretary of State for applications to be "called-in" for his own decision, he retains the right to call in any application (or category of application). However, in practice, these powers are likely to be used sparingly, and generally only in exceptional cases which raise novel or unusual issues, or issues of more than regional significance.

In the event of an application being "called-in" by the Secretary of State, he will consider the application afresh, in the light of evidence submitted by the applicant, the NRA, and any objectors; if necessary, a public local inquiry or a hearing may be held. The Secretary of State would issue his decision in the form of a direction to the NRA (with which it must comply), to refuse the consent, or to grant it, unconditionally or subject to specified conditions.

Consents may be issued without submission of an application form in a situation where trade or sewage effluent has been discharged without consent, or other matter has been discharged in contravention of a relevant prohibition notice or general prohibition, where a similar contravention is likely, and no consent application is forthcoming from the discharger. In such circumstances the NRA may serve on the discharger a consent with conditions specific to the discharge. The discharge is then subject to the conditions of the consent and subsequent breaches will be subject to normal consent enforcement procedures.

4.1.9 Issue of Consent

Once a decision has been reached and all requirements regarding representations or objections to the application have been fulfilled, the consent can be issued.

There is no standard consent format in use across the country although the basic structure, and information contained within discharge consents is very similar in all NRA regions. Standard consent formats and standard clauses for common consent components are being developed. Sections 4.3 and 4.4 provide further details on how consents are currently structured and examples of different types of consent.

4.1.10 Register Provisions

Section 190 of the Act provides for public registers of specified water quality and discharge consent information to be kept and made available for inspection. The detailed requirements are laid out in the Control of Pollution (Registers) Regulations 1989. The Public Registers, compiled and maintained by each NRA region, are structured so that information pertinent to that region can be found by reference to a particular discharge or to a particular location.

Most entries must be made within 28 days of the relevant event, such as the issue, amendment or revocation of a consent. Other entries on the Register include the results of analysis of samples taken by or on behalf of the NRA, for which the deadline for entry is two months from date of sample. Every entry must include the date on which it was first made on the register, and must be retained on the register for at least five years from date of entry, and for as long thereafter as is necessary for the Authority to undertake its functions.

The Public Register includes notices of water quality objectives, applications for consent, consents and conditions and notices served or received in connection with the consent, together with results of analysis of samples of effluent and water. The Register also includes any information relating to authorisations issued under Part 1 of the EPA 90.

The register is available for public inspection at Regional offices, free of charge, with facilities available to obtain copies of register entries for a reasonable fee.

There are variations in the format in which the Register information is available, with some regions using computers for all access and reference, with printouts of entries being available on request; whereas other regions hold the applications as a paper record, with Register consent details and monitoring data on an electronic storage and retrieval system. As progress is made with the NRA's national Water Archive, most Register information will be available on computer.

4.1.11 Appeals

Section 91 of the Water Resources Act details the requirements for appeals by the discharger against the NRA's decision. When the discharger receives the discharge consent, if he is dissatisfied with the conditions specified he may appeal within three months to the Secretary of State. The discharger may also appeal if consent is refused. Determination of the appeal follows a similar approach to that described in Section 4.1.8 for 'called-in' applications, as specified in the Control of Pollution (Consents for Discharges etc.)(Secretary of State Functions) Regulations 1989. Interested parties are given the opportunity to make their case to the Secretary of State, who may hold an inquiry. On determination of the appeal, if the Secretary of State requires the consent to be altered, he issues a direction to the NRA, specifying the new conditions. The amended consent and the direction are placed on the Public Register. Until the appeal is determined the original consent remains in force, and any action arising from it is legally valid.

4.1.12 Charges

Under Sections 131 and 132 of the Water Resources Act 1991 the NRA may recover the costs associated with consented discharges. A scheme of charging for discharges to controlled waters was first introduced in July 1991.

The scheme is for recovery of costs incurred by the NRA in setting consents and monitoring discharges and water quality. It is outlined in Section 10.

4.1.13 Reviews

Schedule 10 of the Water Resources Act places a duty on the NRA to review all consents granted, in particular with regard to the conditions to which they are subject. The conditions defined in a consent normally apply, under current legislation, for a minimum of two years from the issue of the consent before they may be reviewed by the NRA. A review can be undertaken earlier at the request or with the agreement of the discharger, or if necessary to meet international obligations. As a result of the review a decision may be made to revoke or modify a consent. Consents may be revoked if no discharge has been made for the previous twelve months.

Alterations to consent requirements can be achieved in three ways.

- by issue, at the NRA's discretion, of variation notices amending consents;

- by unilateral revocation by the NRA of the existing consent and issue of a new one;

- by a discharger applying for a new formal consent, at the granting of which, the old consent is revoked.

If the first and second routes are used, the discharger does not have to pay an application fee. The latter route would normally apply if there is a material alteration in the discharge.

The provisions for appeal referred in Section 4.1.11 "Appeals", also relate to decisions made by the NRA with regard to consent reviews.

4.2   TECHNICAL CONSIDERATIONS

4.2.1   General

The criteria against which the NRA needs to consider a consent application are the water quality objectives (WQOs) for the waters into which it is proposed to make the discharge. The starting point in the consideration of any discharge consent application is a knowledge of the quality objectives and associated quality standards for the proposed receiving water.

The requirements of the Urban Waste Water Treatment Directive will impose minimum emission standards for sewage effluents and similar industrial effluents. Where environmental needs dictate a higher standard of treatment, then these will prevail.

There are several techniques used to calculate numerical consent conditions, the most common place being the CD methods (combining distributions) which are based on the mass balance equation. Computer models of transport and degradation processes are also used, particularly for discharges into estuaries and coastal waters. There is also increasing use of river catchment modelling, to look at the cumulative impact of effluent discharges on a catchment basis rather than looking separately at individual discharges. Whatever mechanism is used to determine a consent application, the NRA will require appropriate environmental and discharge data on which to base its decisions.

4.2.2   Data requirements

In rivers and streams, the water quality which results downstream of an effluent discharge will be due principally to a combination of:

- the discharge quality;

- the discharge flow;

- the river flow; and

- the upstream river quality.

The onus is on the applicant to provide all relevant data on effluent flow and quality in support of an application. The NRA will normally have river flow and quality data available from its own monitoring. The NRA may require the applicant to obtain additional data if existing data are insufficient to determine the application.

In tidal waters, where water movement patterns are complicated by tidal and weather related effects, the NRA may require substantial amounts of physical, chemical and biological survey data in order to establish the likely impact and fate of contaminants discharged. In most cases the majority of data collection will fall upon the discharger, with the NRA specifying and auditing the quality of the information provided.

4.2.3 Calculation of Discharge Consents for Discharges to Rivers, Estuaries and Coastal Waters

(a) CD-Methods (Combined Distribution Methods)

CD-Methods have been used since the mid 1970's for the calculation of discharge consents and catchment planning. They are based on the mass balance equation.

$$T = \frac{FC + fc}{F + f}$$

Where: T is the concentration of substance downstream of the discharge;

F is the river flow upstream of the discharge;

C is the concentration of substance in the river upstream of the discharge;

f is the flow of the discharge;

c is the concentration of substance in the discharge.

A single application of the mass balance equation cannot be used to calculate the consents needed to meet river targets because the variables in the equation represent values at a particular instant in time. The equation is not true if instantaneous values are replaced by summary statistics such as 95 percentiles. The 95 percentile for T must be calculated from the distribution of T (ie all the possible instantaneous values of T). This distribution can be derived from the distributions for F, f ,C and c, either mathematically (Warn-Brew) or by simulation (Monte-Carlo Simulation).

Detailed information on the calculation of consents is included in Annex 2 of the NRA Guidelines for the Periodic Review (AMP2) (Version 2)(NRA, Dec 1993), and will be included in the NRA Consents Manual.

(b) Computer Models

Computer models are also used in the determination of consent applications, particularly for discharges to estuaries and coastal waters, for which combining distribution methods are not appropriate.

In many cases, predictive models are developed by or on behalf of the discharger in order to provide data to support their consent application. In such cases the NRA's determination of the application would involve an audit of the applicant's model and results, together with any validation data.

The NRA regions are making increasing use of river catchment models. Such models allow the cumulative impact of all effluents discharging to a catchment to be assessed. They contribute to water quality planning for the whole catchment, taking into account the effects of upstream discharges, and abstractions, on the consent conditions required for discharges downstream.

4.2.4 Calculation of Consents for Discharges to Groundwater

The NRA has published its Groundwater Protection Policy which identifies the criteria, based on the vulnerability of aquifers to pollution, which have to be satisfied for discharges to be acceptable for discharge to groundwater.

With regard to the EC Directive on the Protection of Groundwater from certain Dangerous Substances (80/68/EEC), the NRA is actively involved in the determination of appropriate discharge controls. The Authority has been directed by the Secretary of State to implement the provisions of the Directive in carrying out its water pollution control functions.

4.3   CONSENT FORM

4.3.1   Introduction

When the process is finished, and consent granted, a formal document is issued to the applicant; a copy of the consent is also entered onto the public register. The information given on the consent includes administrative information about the application, and any conditions that are to be imposed, in order to comply with water quality requirements and National and European legislation as appropriate.

A typical structure of a consent and the information recorded in it is illustrated in Figure 4.1. Some of the main features of a consent, are summarised in the Sections below.

It is important to note that each consent is uniquely tailored to the particular circumstances of discharge type and composition and receiving water quality requirements. There is no such thing as a "Standard Consent".

4.3.2   Reference to the Act

Prominent on the consent is the Act under which the consent was granted, currently the Water Resources Act 1991. There are still however, valid consents made under previous legislation, such as the Rivers (Prevention of Pollution) Acts 1951-61, Control of Pollution Act 1974 and the Water Act 1989, which can be identified in this way.

4.3.3   Formal clause granting consent

The form of words used to grant the consent summarises the legal status of the NRA in granting the consent, and will name the premises to which the consent applies, and the date of the application. It is important to note that the consent applies to land not to a person, and can be transferred to the new owner or occupier if the premises change ownership.

This section may also include a statement to the effect that the consent does not provide a defence against a charge of pollution in respect of any constituent for which it does not specify consent limits.

4.3.4   Description of Discharge

A general description of the discharge will identify the type of effluent, such as trade effluent, site drainage, final sewage effluent etc, the location of the discharge and the receiving water. This information may be presented within the wording of the formal clause, as a separate section or as one of the conditions of the consent.

4.3.5 Conditions

The conditions of a consent may be presented as a series of numbered paragraphs or may be divided into the following sections:

- General - this will include information such as:

  - date consent comes into force,

  - date following which modifications to the consent may be made.

  - access to and location of the sampling points.

- Outlet - the location and structure of the outfall will be stated, normally by description of location and national grid reference, and the purpose of the outlet declared, ie for the discharge of final effluent, site drainage etc.

- Discharge to watercourse or land - this may include the name of the receiving water and maximum rates of discharge permissible.

- Discharge composition - these conditions will vary according to the source and treatment of the effluent and the assimilative capacity of the receiving waters.

  Numerical limits may be set on water quality determinands such as pH, temperature, BOD, suspended solids and ammoniacal nitrogen. Further specific conditions may also be included in the consent as appropriate depending upon the nature and composition of the effluent.

  Typical descriptive conditions that may be imposed are that the discharge shall contain:

  - no visible evidence of oil or grease;

  - no matter in a concentration that is injurious to fish.

4.3.6 Signature

The consent becomes legally valid following signature by an NRA authorised signatory.

Figure 4.2 Typical Consent Structure and Content

NATIONAL RIVERS AUTHORITY
WATER RESOURCES ACT 1991 - CONSENT TO DISCHARGE

Reference Number: ....

The National Rivers Authority, in pursuance of its powers under the above mentioned Act, HEREBY GIVES CONSENT to the discharge described hereunder subject to the terms and conditions set out below.

Name and Address of Applicant: ....

....

Date of Application: ....

Date of Consent: ....

Description of Discharge: ....

   Type: ....

   From: ....

   To: ....

National Grid Reference of
Discharge Point: ....

This Consent shall not be taken as providing a statutory defence against a charge of pollution in respect of any poisonous, noxious of polluting constituents not specified herein.

Conditions

1 General

2 As to Outlet

3 As to Volume Discharged to the watercourse

4 As to Discharge composition

NRA Region: ....

Address: ....

....

NRA Authorised Signature: ....

4.4   TYPES OF CONSENT

4.4.1   Generic Types

Consents for discharges to controlled waters fall into one of three main types depending upon the nature of the conditions imposed; these are numeric, non-numeric and descriptive consents.

In general, those discharges which have the most potential to exert an undesirable environmental impact are controlled via numeric consent conditions. Numeric consent limits will relate to individual substances or attributes of substances discharged, and will generally specify concentrations and flow.

Typical numeric consent conditions will limit the concentration of a specific substance in the discharge,(eg "No single sample shall contain Dissolved Copper in excess of 5 mg.l^{-1}",) or limit a collective attribute of a group of substances, such as organic oxygen demand from sewage, which may be controlled by a limit on Biochemical Oxygen Demand (BOD).

Numeric limits may be set as absolute values, which are not to be exceeded in any sample, or as percentiles, where a proportion of samples taken over a defined period must not exceed specified concentrations of a suite of substances.

Non-numeric consents are generally used for discharges where the environmental acceptability of the discharge is not readily defined by a limit on concentration. Combined sewer overflows, which should only operate when the flow in the sewer exceeds a defined value as a result of rainfall, are typically subject to such non-numeric consents.

Descriptive consents are normally used only for small discharges which have low potential for adverse environmental effect, typically small sewage treatment works. Descriptive conditions will typically define the effluent treatment plant, and require that it be operated and maintained in accordance with good practice. Environmental conditions describing potential adverse effects in the receiving environment, which operation of the plant must avoid, may also be specified. An example is given in Appendix C.

4.4.2   Trade Effluents

In general, industrial process effluents are subject to absolute limits, as the effluent quality is entirely within the control of the producer.

A typical example of the numeric conditions applied to a trade effluent discharging to a surface watercourse is as follows:

No single sample of the final effluent discharged shall have:

(i)   in excess of 80 milligrams per litre of biochemical oxygen demand (BOD) in the presence of 2.0 milligrams per litre allyl thiourea (ATU) for five days at 20°C;

(ii)   in excess of 120 milligrams per litre of suspended solids (measured after drying for one hour at 105°C);

(iii) in excess of 20 milligrams per litre of ammoniacal nitrogen expressed as nitrogen;

(iv)   a pH value less than 6 or greater than 9.

4.4.3 Municipal sewage effluent

The quality of municipal sewage effluent is not entirely within the control of the sewage treatment companies. The sewage treatment process is itself a biological one, subject to weather impacts, particularly temperature and rainfall, and it is not possible fully to police the sewerage network and discharges made to it, and hence totally control the quality of raw sewage arriving at the sewage treatment works. Additionally, there can be the problems, common to many industries, of vandalism, power and mechanical failure, although the latter two can largely be planned for and ameliorated. Until implementation of the Control of Pollution Act 1974, (COPA), with its requirements for opening to public scrutiny discharge quality information which had hitherto been regarded as confidential, there was a tacit assumption that for sewage discharges, although the consent was expressed as requiring 100% compliance, in practice, if more than 95% of samples met the limit, then that was adequate. In 1985, on implementation of COPA, the Department of the Environment and Welsh Office issued a General Variation Order which changed the legal obligation in all relevant discharge consents to a requirement that the proportion of samples meeting the limits for sanitary determinands, (BOD, Suspended solids and Ammonia), should not be significantly lower than 95%. Assessment of whether or not a sewage treatment works discharge was complying with consent limits would be dependent on a statistically derived "look-up" table which specified the number of failures allowed in any annual set of sample results.

One of the impacts of this change in sewage treatment works consent structure was to weaken the scope for enforcement, as no single exceedance of consent limits would constitute an offence. In its first Water Quality Series report - Discharge Consent and Compliance Policy: A Blueprint for the Future, the NRA recommended that all numeric consents should include absolute limits for all relevant determinands, in order to ensure that the allowed 5% of look-up table exceedances could not legally cause pollution. Accordingly, since publication of the report, the NRA has issued water companies with consents to discharge sewage effluent which include both look-up table and absolute conditions for BOD, Suspended solids and Ammonia. An example of such a consent is given in Appendix C. The water companies have appealed against the terms of these consents to the Secretaries of State.

Industrial effluent components of sewage effluent, such as toxic metals, are subject to absolute concentration and/or load limits in the sewage works discharge consent. These substances are subject to regulation on entry to the sewer by the water companies, and by HMIP in the case of prescribed substances.

Descriptive limits may also be included within numeric consents. Typically, a clause may be included requiring that the effluent discharged shall not be injurious to fish or aquatic life or have any other adverse environmental effect.

4.4.4 Septic Tank Effluent

The Water Resources Act allows the NRA a greater degree of flexibility of control than was available in the earlier legislation. Most septic tanks discharge to land via a soakaway, and for individual small septic tank discharges (less than 5 m3.d^{-1}), it is necessary to decide whether continuing or initial control is needed. If the proposed discharge is close to a groundwater supply source, an absolute prohibition notice will be issued and subsequent consent application refused. If initial control only is required, a conditional prohibition notice will be served requiring that design, siting and maintenance be adequate. If continuing control is needed, because of the risk of the effluent causing an aquifer or surface water problem, or if it exceeds 5 m3.d^{-1}, a consent will be issued. In most cases this will be descriptive.

4.4.5 Transitional Types of Consent

Ideally, all consents would be set in accordance with strictly interpreted environmental criteria. Whilst for most significant discharges this is undoubtedly the case, the changing scope of pollution control legislation has resulted in several transitional forms of consent, some of which are still extant, which legalise discharges but are not necessarily environmentally protective.

Deemed Consents

On implementation of the Control of Pollution Act 1974 in 1985, outstanding consent applications which had been made under the 1961 Rivers (Prevention of Pollution) Act were deemed to have been granted consent pending determination of the application; such cases are referred to as 'deemed consents'. Discharges which were formerly outside the scope of consent controls, notably discharges to coastal waters and pre 1960 discharges to estuaries, were initially exempted from COPA 74 control and in 1987, provided an application for consent had been made, were also deemed to be granted consent for the discharge as described on the application, with determination intended at that time by 1992. Most such discharges are of sewage, and subsequent developments with implementation of EC Bathing Waters and, latterly, Urban Waste Water Treatment Directives, have delayed determination of many consents pending development by the water companies of comprehensive sewerage and sewage treatment schemes to meet Directive requirements.

Time Limited Consents

In the build up to privatisation of the Regional Water Authorities it was recognised that a proportion of sewage works were operating in breach of their consents. The DoE and the Welsh Office varied the individual consents of about 800 sewage works so that their discharges would not be in breach of temporary limits set at the then current levels of performance within the period allowed for capital expenditure by the water companies to improve them. These relaxed consents were subject to time limits, after which the pre-September 1989, or more stringent limits, would apply. These 'time limited' consents also included so called "upper tiers" - a maximum limit set as a multiple of the relaxed 95%ile limit, which it would be an offence to breach at any time. The majority of time limited consents have now lapsed, on completion of remedial works. Some have been extended by the NRA, because of scheme development problems unforseen at the time of privatisation.

Temporary (Schedule) Consents

Also in the run-up to privatisation, as Regional Water Authorities (RWAs) undertook systematic inventories of assets, it was recognised that many combined sewer overflows and emergency overflows from sewage pumping stations were not subject to consent conditions. In order that the new water companies might operate legitimately until the considerable process of setting individual consents could be completed, the Water Act 1989 provided for application for temporary consents to be made to the Secretaries of State. These took the form of schedules listing the limited information available to the RWA regarding these discharges. Often this amounted to a listing of name of discharge, National Grid References, and name of receiving water.

The Secretaries of State then directed the NRA to grant temporary consents for the discharges detailed on the schedule, the temporary consent to have effect until the NRA's final determination of the application. The temporary consent process was aimed at facilitating a prioritised review of the relevant discharges - something that would not have been possible otherwise. There are several thousand such discharges. The NRA is busily reviewing these, but because of the number involved this will take some time with those causing pollution being addressed first.

Additionally, in the run up to privatisation, it was realised that several sewage treatment works were in breach of consented flow conditions and for such works temporary (schedule) consents were issued.

5   NUMBERS OF CONSENTS OF VARIOUS TYPES

In order to assess the overall distribution of numbers and types of consents on the Regional Consents Registers a survey of consents was conducted within the NRA in 1992 and provided a 'snapshot' of the situation, regarding discharge consent numbers and type, at the end of 1991. The quality of the data provided by the regions, for the purpose of the survey, was variable because the data was held in formats which made acquisition and collation difficult.

The data used for this exercise yielded slightly different totals of numbers of discharges than is reported in Section 9 for consent compliance purposes. Nevertheless, as a broad brush assessment for strategic planning of the workload needed to undertake a comprehensive review of consents, the survey delivered an adequate level of information. It must be stressed that the summary statistics presented in this section should be regarded as indicative of distributions, not absolutely accurate statements of numbers and types of discharge.

Figure 5.1 shows the relative distribution of consents across NRA regions. Note that since 1991 Northumbria and Yorkshire regions have merged, as have South West and Wessex regions.

Figure 5.2 shows the distribution of consents by discharge type for England and Wales. Water service company discharges account for about one third of all consents. Of the remainder, about one half are septic tanks, which as the most numerous type of discharge consent, account for approximately one third of the national total. Many septic tank consents, inherited from the former Regional Water Authorities, have been revoked as the discharge did not require continuing control. Those remaining on the Register are mainly required for aquifer or surface water protection purposes.

Figure 5.1 Distribution of Consents Nationally

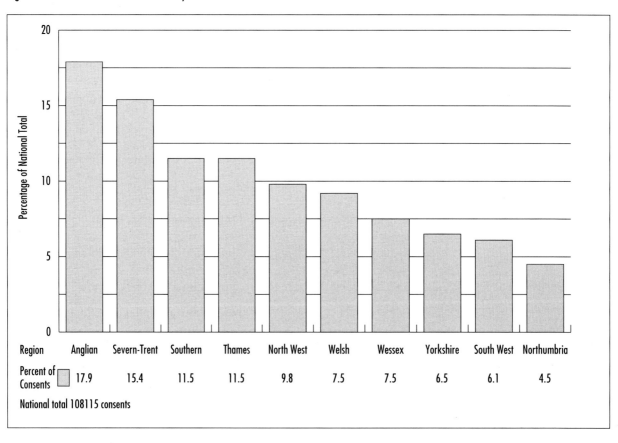

Region	Anglian	Severn-Trent	Southern	Thames	North West	Welsh	Wessex	Yorkshire	South West	Northumbria
Percent of Consents	17.9	15.4	11.5	11.5	9.8	7.5	7.5	6.5	6.1	4.5

National total 108115 consents

Figure 5.2 Consent Type.

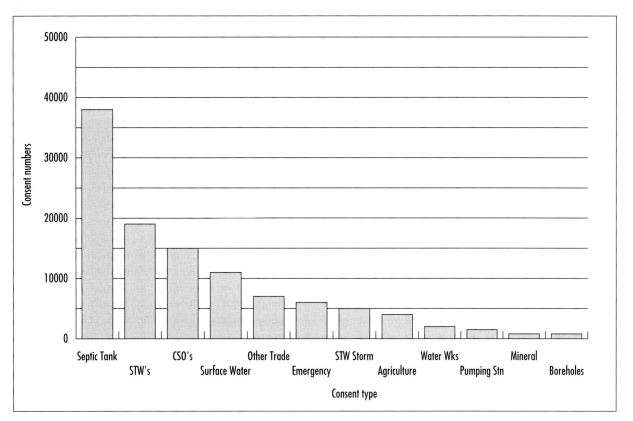

6   KINNERSLEY REPORT

6.1   KEY RECOMMENDATIONS

A report known as 'The Kinnersley Report' - Discharge Consent and Compliance Policy: A Blueprint for the Future; (NRA: Water Quality Series No 1 July 1990) - was produced following the work of a Policy Group on Discharge Consents and Compliance, set up in July 1989, with the following terms of reference: "to review the way in which discharge consents for all discharges to controlled waters are set; the appropriate levels of compliance for different types of discharger; and the way in which compliance with these consents is assessed and monitored."

In all some 33 recommendations were made by the Group, of which three key ones were:

(a) The inclusion of absolute limits for concentrations of determinands which should not be exceeded at any time for those discharges into environmentally sensitive areas for which numeric limits have been set. The introduction of 80 or 50 percentile limits within which the normal operation of the discharge should comply.

(b) For numeric consents, greater restrictions on ammonia should be placed and where previously Biochemical Oxygen Demand (BOD) was used this should be substituted by Total Organic Carbon (TOC); similarly Suspended Solids should be replaced by Turbidity.

(c) The promotion of wider application of automatic and continuous monitoring by dischargers.

A theme common to many of the recommendations was the greater involvement of the discharger in the provision of information to the NRA, both at the time of the application and subsequently, should there be any modifications to the practices engaged in on site that may cause the final effluent to change, either in quality, quantity, rate, nature etc.

6.2   PROGRESS WITH IMPLEMENTATION

Many of the Kinnersley recommendations related to essentially administrative issues which have been adopted generally into working practice and consent processing systems where possible. The NRA is developing a national consent processing system which will replace the existing regional ones. When implemented, the system will ensure that applications are processed in a nationally consistent manner, incorporating the administrative recommendations of the Kinnersley Report. Others have required further consideration, and detailed policy development in order to ensure that a nationally consistent approach is achieved, or require completion of a phase of research and subsequent development prior to a decision to implement them. Developments outside the context of the Kinnersley Report have replaced or amended the significance of some of the recommendations. Such developments include the implementation of the Urban Waste Water Treatment Directive, and the definition of NRA policy affecting the water industry, which has been required for the 5 year cycle of planning inherent in OFWAT's Periodic Review of water industry investment. (NRA Guidelines for AMP-2.) Developments with case law resulting from court decisions have required a reappraisal of some recommendations, and appeal decisions have also provided impetus for refinement of certain consent clauses.

6.2.1   Absolute Limits

The NRA has been concerned to close a legal loophole that potentially allowed sewage treatment works discharges operated by the privatised water companies to cause pollution for five percent of the time. For such discharges, the NRA has adopted a policy that when consents are reviewed absolute limits are imposed in addition to 95 percentile limits. The absolute limits are set as multiples of the 95 percentile.

The water companies have appealed to the Secretary of State against the imposition of absolute limits in new and reviewed sewage works consents. The position regarding determination of the appeals has been further complicated by the Urban Waste Water Directive - which imposes a different monitoring regime upon dischargers than is current practice (as discussed in Section 7). DoE has recently issued policy guidance to the effect that absolute limits should not apply to sewage works with a population equivalent of less than 2,000, and that they should generally only be introduced for other works as part of the implementation programme for the Urban Waste Water Treatment Directive.

6.2.2   Ammonia Limits

The NRA has also pursued a policy of introducing ammonia limits on sewage discharges where it is necessary, in order to protect fish life. The individual discharge limits will vary from area to area, depending on local circumstances and will be based on the relevant WQOs once set. The policy has been applied to works without such limits in a manner intended to enable the operator to achieve them through good operation of existing plant, and without significant additional capital investment.

7 INFLUENCE OF URBAN WASTE WATER TREATMENT DIRECTIVE

7.1 KEY ISSUES AFFECTING DISCHARGE CONTROL POLICY

7.1.1 Introduction

This Directive was adopted in May 1991, with the overall aim of minimising the adverse effects of sewage discharges by requiring treatment to a minimum standard. The Directive will make substantial changes to the way in which consents for sewage treatment works are framed and performance monitored.

The key issues arising from this Directive, that have implications on discharge control policy, are those relating to:

- Identification of Sensitive and Less Sensitive Areas

- Best Technical Knowledge Not Entailing Excessive Cost (BTKNEEC)

- Levels of Treatment

- Emission standards

- Composite samples

Crude sewage discharges from urban areas are currently made to some coastal waters and estuaries. Discharges greater than 10,000 p.e. to coastal waters, and greater than 2,000 p.e. to estuarine waters, will be subject to minimum treatment requirements upon adoption of the Urban Waste Water Treatment Directive.

The Government is producing the necessary regulations and guidance to implement the Directive in consultation with industry and regulators, including the NRA. Much debate has centred on achieving UWWTD requirements without incurring unnecessary costs.

7.1.2 Sensitive Areas and Less Sensitive Areas

The Directive specifies the criteria by which receiving waters may be identified as "sensitive areas" or as "less sensitive areas". The designation of an area will dictate the standard of treatment. Discharges from sewage works with population equivalents (p.e.) greater than 10,000 into sensitive areas will be required to be of a higher standard than secondary treatment. Those to coastal and estuarine waters identified as less sensitive will only require primary treatment. All other waters will be considered "normal", and sewage discharges to such waters will require secondary treatment.

Sensitive areas will be identified where there are:

- natural freshwater lakes, other freshwater bodies, estuaries and coastal waters which are or have potential for becoming eutrophic

- surface freshwaters which are used for potable water supply and which could contain higher concentrations of nitrate than that permissible for waters intended for this use.

Less sensitive areas, which for the purposes of statutory Regulations are termed High Natural Dispersion Areas (HNDAs), are defined as estuaries or coastal waters where the discharge will not have an adverse affect upon the environment. Comprehensive studies will be required to demonstrate that this is the case.

Detailed criteria have been agreed between DoE, NRA and dischargers, to enable identification of the areas to take place. A government consultation paper Methodology for Identifying Sensitive Areas (Urban Waste water Treatment Directive) and Methodology for Designating Vulnerable Zones (Nitrates Directive) was issued in March 1993.

7.1.3   BTKNEEC

The principle of Best Technical Knowledge Not Entailing Excessive Costs (BTKNEEC) is to be applied to the design, construction and maintenance of urban waste water collecting systems. Factors to be taken into consideration in determining BTKNEEC include the volume and characteristics of urban waste water, prevention of leaks from the system and the limitation of pollution of receiving waters due to combined sewer overflows (CSOs).

7.1.4   Levels of Treatment and Emission Standards

The discharges from urban wastewater treatment plants of defined size must meet certain minimum requirements, specified either as concentration limits, or as a percentage reduction in load discharged, of mandatory determinands as set out in the Directive.

For discharges to HNDAs receiving primary treatment, percentage reduction in load of Biochemical Oxygen Demand (BOD) and Suspended Solids (SS) is required.

For discharges to 'normal' waters receiving secondary treatment, concentration limits or percentage reduction may be used to assess compliance with standards for BOD, Chemical Oxygen Demand (COD), and optionally SS.

For discharges to sensitive waters receiving tertiary treatment additional concentration limits for Phosphate and/or Nitrate may apply.

In addition to UWWT Directive requirements, other Directives may necessitate further levels or types of treatment in order to achieve water quality standards set in those Directives.

7.1.5   Composite Samples

The Directive specifies sampling and monitoring requirements which must be conformed with in assessing compliance. The key requirement for the sampling methodology is that flow-proportional or time-based 24 hour samples of effluent, and if necessary influent, are collected at a defined sampling point. Over the specified time period, a number of sub-samples are taken at regular intervals which are combined to constitute the sample that is used for analysis; this is known as a composite sample. This is intended to provide a more representative sample of the overall quality of the effluent than can be obtained from an individual 'spot' sample collected at one instant in time.

The number of samples required will be based upon the size of the treatment plant and must be collected at regular intervals throughout the year.

Compliance will be assessed, in accordance with criteria specified in the Directive, against the Directive's limits. Look-up tables will be used to assess compliance with 95 percentile standards. For any given number of samples taken in a year, a specified maximum number of failures are allowed. Additionally, where concentration limits apply, no sample may breach the upper tier limit set in the Directive, unless it occurs as a result of specified and recorded abnormal operating conditions such as extreme cold.

Composite sampling represents a fundamental change to the UK's historical 'spot-based' practice of sampling for consent compliance. The NRA and water utility companies have commissioned a research and development project to facilitate the change of all sewage consents from spot-based to composite sampling limits. It is also intended that there will be a move towards self-monitoring by the dischargers, with the NRA undertaking less routine monitoring of effluents, and concentrating on audit of the dischargers' monitoring and analytical quality assurance. The transitional steps towards full implementation of this approach are, at the time of writing, under discussion between DoE/WO, NRA and dischargers.

7.2   IMPLEMENTATION

The timetable for the implementation of the Urban Waste Water Directive is outlined in Fig 7.1.

The requirements of the Urban Waste Water Treatment Directive for composite sampling will necessitate detailed revision of discharge consents, but the principles of discharge control outlined in Section 4 will be unaffected. The Directive specifies minimum levels of treatment for sewage discharges. Where the needs of the receiving water require a higher level of treatment, for instance, where there is limited dilution, more restrictive discharge standards will continue to apply.

Fig 7.1: URBAN WASTE WATER TREATMENT DIRECTIVE - IMPLEMENTATION REQUIREMENTS

STW Size	Sensitive Areas			Normal Areas			Less Sensitive Areas	
Population Equivalent	Freshwaters	Estuarine Waters	Coastal Waters	Freshwaters	Estuarine Waters	Coastal Waters	Estuarine Waters	Coastal Waters
	All Areas Identified by end 1993						All Areas Identified by end 1993	
> 15,000	Collection Systems by end 1998 Extra Treatment by end 1998			Collection Systems by end 2000 Secondary Treatment by end 2000			(Category does not apply)	Collection Systems by end 2000 Primary Treatment by end 2000
10-15,000	Collection Systems by end 1998 Extra Treatment by end 1998			Collection Systems by end 2005 Secondary Treatment by end 2005			(Category does not apply)	Collection Systems by end 2005 Primary Treatment by end 2005
2-10,000	Collection Systems by end 2005 Secondary Treatment by end 2005		(Category does not apply)	Collection Systems by end 2005 Secondary Treatment by end 2005		(Category does not apply)	Collection Systems Primary Treatment by end 2005	(Category does not apply)
< 10,000	(Category does not apply)		Appropriate Treatment by end 2005	(Category does not apply)		Appropriate Treatment by end 2005	(Category does not apply)	Appropriate Treatment by end 2005
< 2,000	Appropriate Treatment by end 2005	(Category does not apply)		Appropriate Treatment by end 2005	(Category does not apply)		Appropriate Treatment by end 2005	(Category does not apply)

Sensitive and Less Sensitive Areas to be reviewed at least every four years.
Implementation Programme Reports to be made to the Commission every two years.

▢ Category does not apply.

8 COMPLIANCE ASSESSMENT METHODOLOGY

8.1 COMPLIANCE

A discharge complies when it fully conforms with the limits set in the consent, as measured by the NRA's audit monitoring programme. Irrespective of the numerical value of any limit, compliance is a simple pass/fail measure, which is of direct interest to the NRA and discharger alike.

As regulator the NRA is primarily interested in consent compliance as a measure of dischargers' success in protecting the environment. Provided the consent has been correctly calculated, compliance will ensure that there is no unacceptable adverse effect of the discharge on the receiving water body and its quality goals. Thus the consent will perform its function of preserving quality in the receiving watercourse.

Where a discharge fails to meet its consent various levels of action by the NRA and by the discharger are required, depending on the reason for, and effect of, the non compliance. This is described further in Section 11.

8.2 SEWAGE DISCHARGES

Judging compliance is not always simple. For the WSPLC sewage discharges a complex method of calculating compliance is required in order to take into account the variability of incoming raw sewage, which is outside the direct control of the water undertaker, and seasonal or weather dependent variations in STW performance. Such discharges have consents requiring compliance in approximately 95 percent of samples taken in any twelve month period. Up to 5 percent of samples may legally exceed the 95 percentile limit. To simplify compliance assessment in these cases a 'look-up' table is provided with the consent; this table specifies the number of samples which must comply with the consent per year, based on the total number of samples of the discharge which have been taken in this period.

8.3 DISCHARGE MONITORING

If the consenting process is to be effective as a method of pollution control, it is necessary to monitor the effluent regularly to ensure that the consented parameters are not being exceeded. The discharge is in compliance with its consent provided all, or the necessary number, of the analyses of effluent samples are within the limits set in the consent. Monitoring can be achieved either by direct measurement of a parameter or by sampling and analysis of a number of parameters.

Traditionally, the NRA and predecessors have taken audit samples and had them analysed in order to monitor and assess compliance. Sampling programmes are established to generate the appropriate number of samples over a year, with the date and time of sampling randomised as far as possible in order to make the programme unpredictable to the discharger. Implementation of the UWWT Directive and the introduction of IPC by HMIP for prescribed processes, will place much of the monitoring burden onto the discharger, with the regulator maintaining a specification and quality audit role, both for continuously monitored data and for sample analysis.

8.4 MONITORING WATER QUALITY

Discharge consent levels are set so that, if complied with, the receiving water will comply with its quality objective. In order to confirm that this is so, the receiving water itself must be monitored. In addition, water quality monitoring is carried out for a number of other reasons, such as:

- to check compliance with EC Directives (eg for Quality of Freshwaters Needed to Support Fish Life)

- to collect information (eg as defined under the EC Directive on Exchange of Information on the Quality of Surface Freshwater)

- to comply with international agreements (eg Third North Sea Conference declaration)

- to comply with UK legislation (eg Water Resources Act 1991)

- to provide long term baseline information (harmonised monitoring of UK waters)

- for classification purposes, and temporal assessment of changes in quality.

8.5   CONSENT COMPLIANCE ASSESSMENT

At present, the NRA is reviewing its policy with respect to consent compliance assessment. This review is based around two fundamental criteria in the sampling and assessment process as follows:

- a consideration of the merits of composite versus spot samples for judging consent compliance

- how sampling and analytical data is processed when compliance is being assessed.

It is probable that the NRA's discharge consent compliance assessment methodology will continue to develop over the next few years.

A discharge consent for a Water Company sewage treatment works will normally specify one or both of two standards:

- absolute standards which must be complied with at all times

- percentile standards with compliance assessed using a look-up table.

Other types of consent conditions are also possible. In particular, some consents specify "load" conditions, particularly for dangerous substances, which limit the mass of such substances that may be discharged within unit time. Load conditions are particularly important in controlling the input of substances which may have relatively low acute toxicity, but which exert chronic effect through accumulation in the environment. A consequence of this type of condition, is that the flow of the discharge must be measured at the same time as the concentration of substance in the discharge in order that load compliance may be judged.

Assessment of compliance with absolute standards is simply a matter of arithmetic comparison of the measured value and the consent limit for each consented parameter, in any monitoring sample. If any parameter is outside its permitted limit the discharge as a whole is deemed to have failed its consent for that sample, and for the reporting period, typically a quarter or year.

Individual monitoring sample results for consents which include look-up tables are judged for compliance in any rolling year (12 month period) by cross-checking on the table the number of samples which must pass given the total number of samples taken during that rolling year.

As well as numerical quality conditions, a consent can specify many other conditions which the discharger must comply with. Examples are:

- that the flow of the discharge be continuously measured and recorded in accordance with NRA specification;

- that records are kept of, for example, plant maintenance;

- that on-line monitoring must be carried out for particular parameters, and summary data be derived and made available for inspection, or be reported.

In these and other such cases, non-compliance is failure to carry out the tasks specified in the consent documents, and may lead to enforcement action for breach of consent conditions.

On-line, or continuous, receiving water monitoring is also carried out, in some NRA regions, to check the effect of discharges on compliance with river quality standards. A further development of continuous monitoring for compliance assessment has been the subject of an NRA research study which is now entering its development stage. This is the development of an automatic monitor and sampler which can take formal, and legally admissible, tri-partite samples once continuous monitoring has shown the discharge to be in breach of its consent. This will make the task of identifying, and if necessary prosecuting, polluters considerably easier. A successful prosecution has already been mounted and further field trials are under way.

9   COMPLIANCE STATISTICS

9.1   INTRODUCTION

Because of the diversity of NRA regions, in terms of water quality problems, consents, monitoring approach and organisational structure, the reporting of consent compliance to an adequate, nationally consistent degree of accuracy has proved very difficult. The data presented below is as accurate as the recording systems then in place allowed. It may not be possible to directly compare the 1990-92 data with subsequent data sets, as from January 1993, the regions adopted a tightly specified consent compliance reporting programme, which, along with monitoring changes will generate more meaningful and consistent summary statistics.

The NRA will publish annual summary reports on consent compliance, covering a calendar year's data. The 1993 data will be published, later in 1994, after publication of this report. Regions may also report consent compliance, on a more specific, local basis to Rivers Advisory Committees, and this information will normally be available from the relevant regional office, on request, to interested members of the public.

Consent Compliance can only be reported where there has been monitoring of a discharge. Many discharges are not directly monitored because of their insignificant nature or low risk of causing pollution. The receiving waters are subject to monitoring programmes and, in the event of a deterioration being detected, investigational sampling may be undertaken, of discharges, in order to establish the cause and ensure that the position is rectified.

9.2   COMPLIANCE DATA 1990-92

Summary statistics on consents and compliance derived from regional returns for 1990-92 calendar years are presented in:

Figure 9.1, Number of Consented Discharges,

Figure 9.2, WSPLC Sewage Discharges Compliance, and

Figure 9.3, Industrial Discharges Percent Compliance.

Since 1992 there have been some organisational changes within the NRA, and following the merging of Northumbria and Yorkshire regions, and South West and Wessex regions, there are now eight regions. Future reports on compliance will reflect the new structure.

Because of the inter-regional variability, both in terms of consents structure and monitoring, it is not practical to undertake a rigorous analysis of the underlying data for that period.

It is important to bear in mind that, historically, consenting procedures were different in the predecessor bodies to the NRA, and to a large extent this legacy remains on the consents Register. Comparative assessment of consent compliance between, and within regions, depends on the stringency with which individual consents were set. Regional variations in consent setting and monitoring are being addressed through introduction of a common discharge monitoring policy, and production of specific policies and guidance on consents issues. These will be included in the Discharge Consents Manual, which is currently being drafted as an updatable document, for issue later in 1994.

Whilst every effort has been made to ensure that Figures 9.1 to 9.3 are correct, they should be regarded as illustrative of the overall position, and not as definitive data sets for direct inter- and intra- regional comparison.

9.2.1 Fig 9.1: Number of Consented Discharges

The substantial reduction in numbers of consents from 1990 reflect the rigorous 'pruning' of obsolete consents from the register, particularly in Anglian, Severn Trent and North West Regions, as a result of implementation of the Charging for Discharges Scheme.

Most Regions experienced approximately equal numbers of revocations of consents, (mostly resulting from introduction of charging for discharges,) and determination of new discharge consents over the period.

Figure 9.1 Number of Consented Discharges

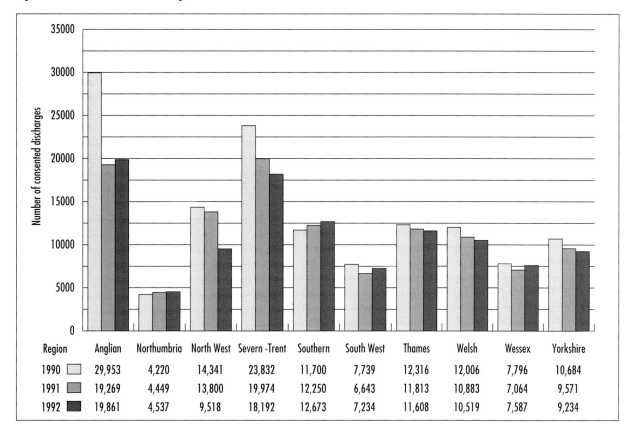

Region	Anglian	Northumbria	North West	Severn -Trent	Southern	South West	Thames	Welsh	Wessex	Yorkshire
1990	29,953	4,220	14,341	23,832	11,700	7,739	12,316	12,006	7,796	10,684
1991	19,269	4,449	13,800	19,974	12,250	6,643	11,813	10,883	7,064	9,571
1992	19,861	4,537	9,518	18,192	12,673	7,234	11,608	10,519	7,587	9,234

9.2.2 Fig 9.2: WSPLC Sewage Works Discharges 1990-92 - Percent Consent Compliance

Compliance was assessed against all elements of WSPLC STW consents, ie 95%ile look-up table compliance and, where set, absolute limit compliance.

Generally WSPLC sewage works compliance has been very good. Most Regions showed an increase in compliance over the period 1990 to 1992 with national compliance for WSPLC sewage works rising from 90% in 1990 to 95% in 1992. However, it must be pointed out that a substantial number of consents that the NRA inherited were not set on a 'river needs' basis.

Consent compliance in South West Region is anomalously low. This is, at least in part, due to particularly stringent consent conditions set for STWs, prior to establishment of the NRA, coupled with the particular problems experienced in the South West, notably short, flashy rivers and seasonal impact of a large tourist population.

Figure 9.2 WSPLC Sewage Works Discharges 1990-92 - Percent Consent Compliance

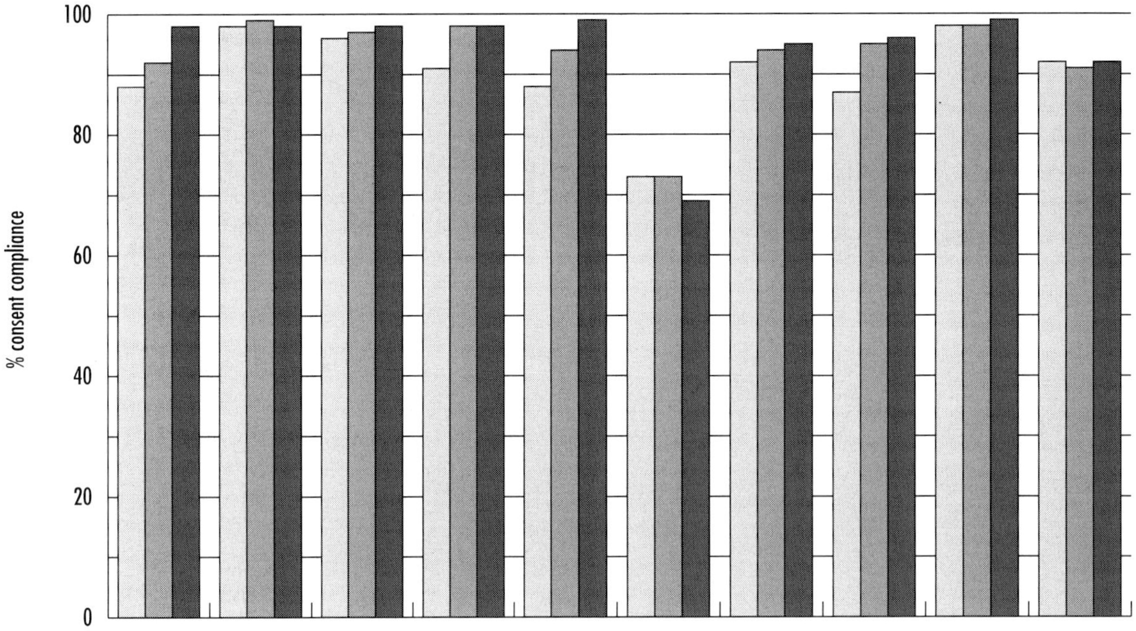

Region	Anglian	Northumbria	North West	Severn -Trent	Southern	South West	Thames	Welsh	Wessex	Yorkshire
1990	88	98	96	91	88	73	92	87	98	92
1991	92	99	97	98	94	73	94	95	98	91
1992	98	98	98	98	99	69	95	96	99	92

9.2.3   Fig 9.3: Industrial Discharges 1990-92 - Percent Consent Compliance

The position with regard to industrial consent compliance appears far less satisfactory than is apparent for STWs, with only about two thirds of industrial discharges fully complying with consent conditions throughout the period.

In part, the nature of industrial discharge consents can be expected to generate lower annual compliance statistics as only absolute limits are used for industrial consents. Risk of failure of any discharge therefore increases as the frequency of consent monitoring increases, and as the number of consented determinands increases, as any single determinand failure within the reporting period results in overall failure of the discharge to comply with its consent. Industrial discharge consents are set with stringent limits, which should, in general, be achievable if the treatment plant is run properly.

The NRA's approach to dealing with issues of non-compliance with consents is outlined in Section 11.

Figure 9.3 Industrial Discharges 1990-92 - Percent Consent Compliance

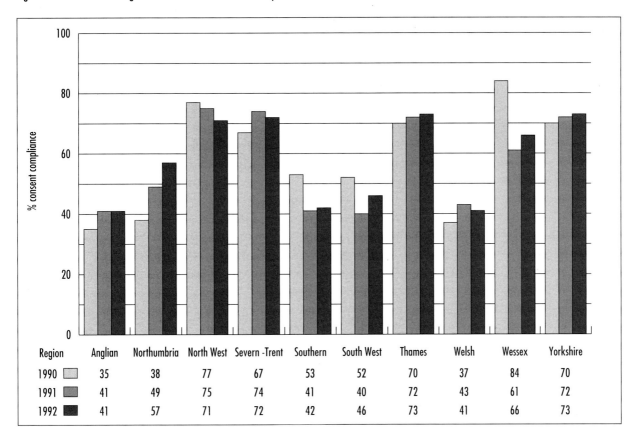

Region	Anglian	Northumbria	North West	Severn -Trent	Southern	South West	Thames	Welsh	Wessex	Yorkshire
1990	35	38	77	67	53	52	70	37	84	70
1991	41	49	75	74	41	40	72	43	61	72
1992	41	57	71	72	42	46	73	41	66	73

10   CHARGING FOR DISCHARGES

10.1   THE CHARGING SCHEME

The Water Act 1989, and subsequently the Water Resources Act 1991, give the NRA a power to recover the costs of administering the issue of consents, and monitoring compliance with them. Previously such work was funded by the Government through Grant-in-Aid from the Department of the Environment. The effect is to transfer the costs to those who make the discharges rather than the general tax-payer.

In July 1991 a scheme of charging for discharges was introduced which is applicable to consented discharges to controlled waters in England and Wales. Minor amendments to the scheme were introduced in April 1994. The most significant of these relate to the weighting factors for volume and content, which have been revised to reflect more accurately the frequency and cost of monitoring. Details of the revision are provided in an NRA Consultation Document issued in September 1993.

It has been the intention of the Government that, after introduction of the initial charging scheme, the charges levied should rise progressively so that the full costs of all consents related activities can be recovered. This is now the case, and enables the NRA to progress with consents issues associated with implementing water quality objectives and pollution control measures. The NRA still requires income from the Government in order to carry out more general environmental monitoring, and pollution prevention and control work that is not directly related to discharges.

10.2   FINANCIAL INCENTIVES

Dischargers have a clear interest in ensuring that, as a minimum, they comply with consent limits at all times, as to fail to do so runs the risk of detection and enforcement action by the NRA. However, in most cases, it is in the discharger's best interest to perform well within the consent limits, in order to minimise the waste of valuable process materials within the effluent stream. The NRA, in conjunction with the Department of Trade and Industry, DoE and HMIP, is actively pursuing a policy of promotion of waste minimisation in industries within selected river catchments. Some remarkable savings have been made by participating industries. Quite simply, millions of pounds worth of potential profit are being poured down the drain by many industrial sectors.

The introduction of the current Charging for Discharges Scheme, undoubtedly gave many dischargers an incentive to reduce the quantity and improve the quality of effluent discharged, in order to minimise their bills. However, this can be looked upon as a one-off incentive, as the scheme is specifically aimed at NRA cost recovery, and does not contain a mechanism for introduction of true incentives.

Incentive charging schemes are being used successfully in other countries and are being considered, for possible future implementation in the UK. Currently, the law does not allow for incentive charging whereby improvements in a dischargers performance are rewarded by reduced charges whilst deterioration attracts higher charges. Financial penalties currently arise only if prosecutions have successfully been brought, or if clean up action has been carried out in response to an unauthorised discharge, or civil action is undertaken by injured parties to sue for damages. (Where a pollution incident occurs, whether or not involving a discharge which is subject to consent conditions, the NRA seeks to recover its clean up and legal costs from the discharger.)

The revised charging scheme allows for charges to be reduced where, with the agreement of the NRA, the discharger undertakes the monitoring of his discharge and reports the information to the NRA. For this option to work the discharger must satisfy the NRA that his sampling and analytical facilities at least match those of the NRA. At the time of writing, the detailed consent requirements for such discharges have not been resolved, and it is unlikely that any such arrangements will be made before the 95/96 charging year.

10.3   TYPES OF CHARGES

There are two elements of charge associated with discharge consents: an application charge and an annual charge. Charge rates are set annually.

Application Charge

The application charge is payable when a new consent is applied for or an application is made for an existing one to be revised. The details of these charges are provided with the consent application form.

Annual Charge

The annual charge is based upon three factors related to the discharge consent, namely:

● volume,

● content,

● receiving water,

together with the financial factor, which is set annually.

10.4   CALCULATION OF ANNUAL CHARGES

To simplify the procedure for calculating the annual charge for a consent, the parameters that describe the volume, content and receiving water, have been divided into bands (Table 10.1). Each band has been ascribed a value or factor which reflects the cost associated with that band. Table 10.1 summarises the charging bands as proposed in the NRA's Consultation Paper for the amended scheme to take effect from April 1994.

The factor reflects the costs of monitoring in each case. For example the weighting factor for volume takes into account that although larger volumes do cost more to monitor, the costs do not increase in direct proportion to volume. In the case of content the factors allow for higher analytical costs, together with the greater complexity of monitoring required for the more significant effluents. The weighting factors for receiving waters takes into account the varied frequency of both discharge and related environmental monitoring, and particularly the cost and complexity of the monitoring. The high factor ascribed to discharges to an estuary reflects the sampling and analytical problems associated with this environment.

Thus, a discharge of an effluent which has a large consented volume, contains pesticides, and is made to an estuary will attract a high charge.

The annual charge is calculated by multiplying the appropriate factors for the discharge to give the number of chargeable units:

volume factor x content factor x receiving water factor = chargeable units

The number of units is then multiplied by the current years 'financial factor' to produce the annual charge.

Table 10.1 - Consent Charging Bands

VOLUME	
Cubic metres per day	Factor
0-5	0.3
> 5 - 20	0.5
> 20 - 100	1.0
> 100 - 1,000	2.0
> 1,000 - 10,000	3.0
> 10,000 - 50,000	5.0
> 50,000 - 150,000	9.0
> 150,000	14.0
CONTENT	
Band	Factor
A - Complex organic, pesticides	14.0
B - Potentially toxic, metals etc	5.0
C - Organic sewage/Trade effluent	3.0
D - General trade effluent	2.0
E - Site drainage	1.0
F - Low environmental effect	0.5
G - Minimal environmental effect	0.3
RECEIVING WATER	
	Factor
Ground	0.5
Coastal	0.8
Surface	1.0
Estuarial	1.5

The following example illustrates how the charging scheme works.

Example: A consent allows for a discharge of 7,000 cubic metres per day of general trade effluent to a river.

Substituting the appropriate factors in the above equation:

3.0 x 2.0 x 1.0 = 6.0 chargeable units

Multiply the number of units by the 'financial factor' of £389 for 1992/93 gives an annual charge of £2334.

Further details of the charging scheme can be found in a leaflet entitled 'Annual Charges - Discharges to Controlled Water' available from the NRA.

11 ENFORCEMENT FOR NON-COMPLIANCE

11.1 INTRODUCTION

The NRA maintains a large monitoring programme to check on the compliance of dischargers with their consents. For the most part this requires the taking of samples which are then analysed for the relevant determinands. Unless the effluent is visually of poor quality, or on-site instrumentation indicates a process failure, or there is environmental evidence of adverse effect, an assessment of compliance with consent conditions cannot be made until the effluent sample analysis has been undertaken.

If routine monitoring indicates breach of consent conditions or a deterioration in effluent quality over time that may lead to a breach, the NRA takes appropriate action to ensure that the discharge returns to compliance. The action will depend on the severity of the impact and the reason for non-compliance; it may be to prosecute, or to issue a warning to the discharger requiring improvements within a set timescale, which if not achieved could lead to prosecution. It is the NRA's remit to be a strong and consistent enforcement agency, and it intends to be fair and equitable to all dischargers, taking a firm line on breaches of consent.

11.2 ENFORCEMENT AND EVIDENCE

Having taken account of the appropriate legislation, granted a consent and stipulated certain conditions, the NRA monitors the discharge and receiving water, in accordance with its monitoring programme. The monitoring data is used to assess the compliance of that discharge with the consent conditions as described in Section 9. In the event of an offence occurring and the relevant evidence being available, the NRA may proceed with a prosecution for an offence under Section 85 of the Water Resources Act. Third parties may also initiate a prosecution if the NRA decides not to proceed.

The analytical result of any sample is not admissible as evidence in any subsequent legal proceedings unless the sample has been taken in a manner described in Section 209 of the Act. For a sample to be admissible, the person who took the sample must:

(1) Notify the occupier of the land of his intention to have the sample analysed;

(2) There and then divide the sample into three parts, and place each in a separate container, which is then sealed and labelled; and

(3) Deliver one part of the sample to the occupier of the land, retain one part and submit the third for analysis.

This is known as the tripartite procedure for sampling and the above steps must be strictly adhered to, or complied with as soon as reasonably practicable after the sample was taken, if the results of analysis of the samples are to be admissible as evidence of an offence.

11.3 REASONS FOR NON COMPLIANCE

There are two basic reasons for non compliance - either the treatment plant is inadequate or the operation and management of the plant is deficient. On investigation it may be found that both apply in a particular case.

11.4 PLANT INADEQUACY

Where a treatment plant is inadequate to achieve consent requirements, two scenarios may apply.

11.4.1

If the discharger has an existing programme of remedial works already agreed with the NRA, the NRA will not normally prosecute for breach of consent pending completion provided that:

- the non compliance does not result from poor management or from problems unconnected with the proposed remedial works; and

- every effort, including temporary works, is being made in the interim to comply; and

- the breach is not of such magnitude as to cause persistent pollution or to justify criminal proceedings in the public interest.

11.4.2

If the discharger does not, at the time of non compliance, have a planned programme for improvement, he will be notified that he is liable to prosecution if the discharge remains non-compliant, but will be given the opportunity to produce an action plan, to NRA agreement, to meet existing consent conditions, (unless there is a clear need for the consent to be revised, such as current failure of the receiving water to meet its class, in which case the plan will need to address the new consent requirements). Provided the plan is adhered to, the NRA will not prosecute, subject to the same qualifications as in 11.4.1.

11.5 MANAGEMENT FAILURE

There are two approaches to enforcement of consents where there has been a management failure to operate the treatment plant property, depending on the structure of the consent.

11.5.1 Absolute Limits

Where the consent specifies absolute limits, (ie industrial effluents, private sewage treatment works, Water Company STWs with upper tiers on sanitary determinands and absolute limits on trade effluent constituents,) the discharger will be warned, by letter, of the non-compliance that has been detected by routine monitoring, and asked to explain the reason. The discharger will be notified that the next sample will be taken in the tripartite manner, and if non compliant, may lead to prosecution.

In such cases a decision to prosecute on the basis of individual tripartite samples, will be dependent upon a number of factors, including the degree of impact of the discharge on the environment, and the attitude of the discharger to improving treatment plant management.

11.5.2 Look-up Table

Where the consent includes a look-up table, (ie Water Company STWs,) enforcement action leading to prosecution is considerably more resource intensive and time consuming.

When an individual sample fails the numerical look-up table limit, the discharger is warned of the exceedance and asked to provide a explanation. If it is evident that this is not likely to be a one-off event, routine sampling will continue on a tripartite basis until sufficient evidence has been collected to form the basis of a prosecution. This is normally at least 12 months data as, in order to be eligible as evidence, all samples, including compliant ones, have to be taken on a tripartite basis, and the look-up table applies to twelve consecutive months of data.

In many cases initiation of routine tripartite sampling has provided sufficient stimulus to generate the required management improvement, and works have quickly been restored to compliance, in which case normal monitoring has been recommenced.

11.6   Enforcement statistics

In many cases of breach of consent conditions, the breach is transitory, not evident at the time of initial sampling, and is not repeated when subsequent tripartite samples are taken. It follows, therefore, that even if it were NRA policy to prosecute for all breaches of consent, it would not be possible to do so in all cases. As outlined in sections 11.4 and 11.5, where evidence is available, the NRA may choose not to prosecute in certain circumstances, in particular with regard to the impact of the discharge, and attitude of the discharger to improved performance.

Statistics for 1990 to 1992 for successful prosecutions where there was a breach of consent conditions are presented in Figure 11.1. Cases may have been taken under Section 85 (1) of the Water Resources Act - for causing or permitting pollution, or under Section 85 (3) - for discharging effluent, or Section 85(6) - for contravening consent conditions.

Figure 11.1 Successful Prosecutions 1990-92 - Arising from Breach of Consent Conditions

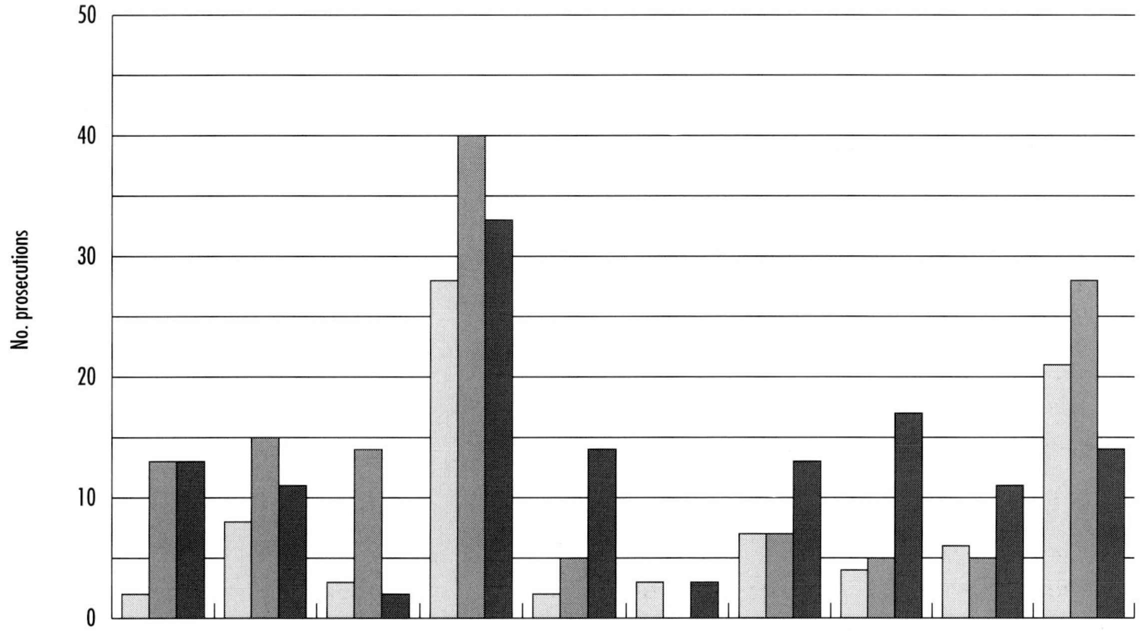

Region	Anglian	Northumbria	North West	Severn -Trent	Southern	South West	Thames	Welsh	Wessex	Yorkshire
1990	2	8	3	28	2	3	7	4	6	21
1991	13	15	14	40	5	0	7	5	5	28
1992	13	11	2	33	14	3	13	17	11	14

REFERENCES

Howarth W, *Water Pollution Law,* Shaws and Sons Ltd, 1988.

Howarth W, *The Law of the National Rivers Authority,* NRA, 1990.

Water and the Environment - The Implementation of Part 2 of the Control of Pollution Act 1974, Joint Circular from the Department of the Environment and Welsh Office, (Circular 17/84 - DoE, Circular 35/84 - WO,) July 1984.

River Quality: The Governments Proposals, DoE, December 1992.

Planning the Achievement of River Quality Objectives - Calculation of Consents, NRA, 1992.

Proposals for Statutory Water Quality Objectives, NRA 1991.

Water Pollution Incidents in England and Wales, NRA, 1990, 1991, 1992.

Policy and Practice for the Protection of Groundwater, NRA 1992.

Guidelines for AMP(2) Periodic Review, Version 2, NRA, Dec 1993.

Charging for Discharges - Consultation Document, NRA Sept 1993.

GLOSSARY

Absolute limit

A consent limit not to be exceeded in any circumstance.

BOD

Biochemical oxygen demand - an index of water pollution which represents the content of biochemically degradable substances to the water or effluent sample.

COD

Chemical Oxygen Demand - an index of water pollution which represents the total amount of oxygen that could be consumed by oxidative degradation of the effluent or water sample.

Consent

A legal authorisation to discharge effluent, which, if complied with, acts as a defence against prosecution for pollution offences.

Controlled waters

Controlled waters include all rivers, canals, streams, brooks and drainage ditches, lakes and reservoirs, estuaries, and coastal waters, and groundwater. Small ponds and reservoirs which do not themselves feed other rivers or watercourses are not included within the definition of 'controlled waters' unless the Secretary of State defines them as such - which he has done in the case of water supply reservoirs in the Controlled Waters (Lakes & Ponds) Order 1989.

'Controlled waters' are defined (Section 104) as:

(a) Territorial waters, ie waters extending to three miles from the baselines from which territorial seas adjacent to England and Wales are measured.

(b) Coastal waters, ie those waters within the area landward of the territorial sea baseline to the limit of the highest tide or, in the case of estuaries, to the freshwater limit of the river or watercourse.

This definition of coastal waters also includes docks and associated areas.

(c) Inland freshwaters, including reservoirs, ponds and lakes, and rivers or watercourses above the freshwater limit.

(d) Ground waters contained in underground strata.

Descriptive consent

A consent which qualitatively describes the type of treatment or polluting effect rather than setting numerical limits, normally used for small sewage works.

Determinand

A substance , specified in a consent or quality standard, that is determined by analysis.

Domestic waste

Waste waters arising predominantly from residential premises and services.

Effluent

Any water borne contaminating substances that enters the environment, normally waste waters from processes.

Effluent discharge

The release of a volume of effluent to the environment.

EQO	Environmental Quality Objective. A formal statement of the desired use of a particular water, such as potable abstraction or fishing. The term is also used, specifically in the context of EC Directives, to mean a concentration of a particular dangerous substance in receiving water with which compliance must be demonstrated.
EQS	Environmental Quality Standard - a standard, normally a concentration, which, prescribes the level of a substance in the environment, which may not be exceeded.
Estuary	Tidal mouth of a river, often subject to large temporal and spatial changes in salinity as freshwater mixes with sea water.
Eutrophic	Nutrient enriched, and leading to undesirable plant or bacterial growth.
Fresh-water limit	Defines the limit between coastal and inland waters
Groundwaters	Waters below the surface of the ground contained within underground rocks, includes well, boreholes and mine waters.
Industrial waste water	Waste waters arising from trade or industrial premises.
Inland waters	See 'Controlled Waters'.
Non numeric consent	A consent that does not set numeric quality limits, but relies on specification of a numeric process, variable such as flow in order to achieve the required degree of environmental protection.
Numeric consent	A consent which sets numeric limits on the quality of the discharge.
Package sewage treatment plants	A small, self-contained commercially available sewage treatment facility used to treat effluent from private houses and small communities.
Percentile limit	A limit which requires the quality of the discharge to be maintained below a numerical value for a certain percentage of time.
Population Equivalent (p.e.)	1 p.e. (population equivalent) means the organic biodegradable load having a 5 day biochemical oxygen demand (BOD5) of 60 g of oxygen per day.
Primary treatment	The physical treatment of sewage effluent, usually settlement, to remove gross solids and reduce suspended solids by about 50%, and BOD by about 20%.
Red List	A list, resulting from the North Sea Conference, of 23 potentially dangerous substances, which are subject to control, to reduce their input from the UK to controlled waters.
Relevant Territorial Waters	See 'Controlled Waters'.
Secondary treatment	Biological treatment and secondary settlement of sewage effluent, normally following primary treatment, capable of producing a substantial reduction in BOD and suspended solids.

Sewage	Wastewater carried in sewers may be a combination of liquid or water carried domestic, municipal and industrial wastes, augmented by such groundwater, surface water and storm water as may be present.
Sewerage	System of foul or surface water drainage by sewers, including any overflows.
Tertiary treatment	Any treatment following secondary treatment which produces a high sewage effluent quality by means of, for instance, grass plots, micro-strainers or nutrient removal.
TOC	Total organic carbon - all the organic carbon present in an effluent or water sample.
Tripartite sample	A sampling method for use when a sample analysis is needed as evidence, where a sample is divided into three parts immediately after having been taken, one of which is sent for analysis, one to the discharger and one is kept by the NRA for production in court.
Turbidity	The opacity of a liquid to light due to particles in suspension. May be used as an on-line measure of suspended solids.
Urban waste water	Domestic waste water or a combination of domestic waste water, industrial waste water and/or surface run-off.
WQO	Water Quality Objective - a set of requirements to be met to achieve specified water quality standards.

APPENDIX A: Resumé of Earlier Pollution Control Legislation

A.1.   PUBLIC HEALTH ACTS 1848 - 1872

The second half of the 19th century saw the evolution of legislation to treat sewage and other effluent prior to discharge as a necessity for eliminating waterborne diseases.

The 1848 Public Health Act established, for the first time, provisions for improving sanitary conditions in England and Wales under the control of a single management body. It placed responsibility for the provision and maintenance of sewers and sewerage works on Local Boards of Health. It also made provisions for water supply, and more importantly, for the protection of that supply.

Subsequent amendments to the Public Health legislation were made over the following 25 years. Of these, one of the most significant was an amendment in 1861 which required the Local Boards of Health not simply to dispose of sewage or wastewater but to treat it so that the final effluent discharged would not affect or deteriorate the receiving water-course.

Sewage treatment resulted in the production of quantities of sewage sludge at the works which posed a further problem for storage and/or disposal. Application of sewage sludge to land was facilitated by the 1865 Sewage Utilisation Act. Disposal for agricultural purposes was advantageous in two ways; it reduced storage requirements as well as being a good soil conditioner and fertiliser.

The 1872 Public Health Act defined urban and rural sanitary districts and placed the responsibility for sewage disposal on the Local Authorities.

A.2.   SALMON FISHERIES ACT 1861

In 1860 a Commissioners' report on salmon fisheries, highlighted the decline in productivity of salmon fisheries in rivers polluted by mines, factories and gasworks effluent. The 1861 Salmon Fisheries Act, which prohibited pollution of salmon waters with substances harmful to fish, was a consequence of this report. This Act had limited effectiveness and was replaced a few years later by the more comprehensive Rivers Pollution Prevention Act 1876.

A.3.   RIVERS POLLUTION PREVENTION ACT 1876

Two Royal Commissions, established in 1865 and 1868, investigated sources of river pollution and identified preventative actions. Their conclusions were that legislation was required to control discharges of sewage, mining and industrial wastes. The recommendations were wide ranging and can be summarised in three main points. First to prohibit any discharge of solid matter into a surface watercourse. Secondly, to establish quality criteria with which the discharge should comply; and thirdly, that manufacturers should be allowed to discharge their effluent into the town sewers. At the time these recommendations were considered too radical to be implemented, although they form the basis of current legislation governing effluent discharges. The legislation that was introduced, as a result of the findings of the Commission, took the form of the Rivers Pollution Prevention Act, 1876, which was concerned exclusively with pollution prevention.

The effectiveness of this Act in preventing river pollution was poor, primarily as a result of limitations in its enforcement. The Act was finally repealed in 1951.

A.4.   PUBLIC HEALTH ACT 1936

Over the period 1890 to 1936 a number of amendments were made to public health legislation. The more significant of these related to the discharge into a public sewer of waste that would constitute a nuisance, or was dangerous or hazardous. All these amendments were consolidated into the 1936 Public Health Act.

A.5.   RIVER BOARDS ACT 1948

The River Boards Act 1948 made provisions for the restructuring of the administrative aspects of river pollution control, which had been instrumental in the relative ineffectiveness of earlier water pollution legislation. The solution adopted was the formation of 32 River Boards each with responsibility for pollution control of an entire catchment area. This structure replaced the previous multiplicity of controlling bodies and fragmented areas of responsibility within the local government framework.

A further major innovation within the Rivers Boards Act was the empowerment of the River Boards to sample any effluent discharging into any inland waters under their jurisdiction.

A.6.   RIVERS (PREVENTION OF POLLUTION) ACT 1951

The 1876 Act was replaced by the Rivers (Prevention of Pollution) Act 1951. The prime objective of this Act was to prohibit the use of surface watercourses for the disposal of polluting material. This applied equally to sewage and industrial pollution and ended previous distinctions made between solid and liquid pollution. An offence under this Act was subject to a number of qualifications. In particular, provisions were made for new trade or sewage effluent discharges to be licensed, ie the discharge was allowed with the consent of the appropriate River Board. Existing discharges did not require a consent. The consent would be permitted subject to conditions relating to the point of discharge to a watercourse, the structure of the outlet, the nature, composition, temperature, volume and rate of the discharge.

A.7.   CLEAN RIVERS (ESTUARIES AND TIDAL WATERS) ACT 1960

This Act was essentially an extension of the Rivers (Prevention of Pollution) Act 1951. It empowered the River Boards to deal with new discharges into specified estuaries and tidal waters in the same way as for discharges into inland watercourses.

A.8.   RIVERS (PREVENTION OF POLLUTION) ACT 1961

This Act brought about a significant reform in the control of discharge of trade and sewage effluents to fresh waters, which, because they were in existence before the 1951 Act, were not previously subject to consents. These discharges now became subject to consents for which the application had to quantify the nature, composition, temperature, maximum volume and flow rate of the effluent. Existing discharges to estuaries and coastal waters remained outside discharge consent requirements.

A.9.   THE WATER RESOURCES ACT 1963

Reports from the Central Advisory Water Committee, which had been created to examine increasing demands for water lead to introduction of the Water Resources Act 1963. This encompassed many aspects of the conservation and use of water resources, including control of abstractions for water supplies, together with new powers and responsibilities relating to river pollution control.

This Act replaced the 32 River Boards, created under the River Boards Act 1948, with 27 River Authorities. Amongst the pollution control measures introduced was the control of discharges into underground strata by a consent procedure. Effectively underground discharges became subject to a similar consent system as applied to surface water discharges.

The Act also empowered the River Authorities to sample any effluent discharging into inland or coastal waters in the authority area. The sampling procedures were subject to the same tripartite division of sample approach, as defined in the 1948 Act, for them to be admissible as evidence of an offence.

Further new powers under this Act allowed the River Authorities to undertake emergency measures, to mitigate or remedy pollution, in response to pollution incidents; and to make byelaws to protect water resources.

A.10.   THE WATER ACT 1973

The objectives of this Act were to resolve existing conflicts of interest between the various authorities concerned in water conservation and pollution prevention and to formulate a comprehensive water management plan on a river basin level. To achieve these objectives ten, multi-purpose, Regional Water Authorities were created. These Authorities took over responsibility for pollution control, resource conservation, fisheries and flood defence from the 27 River Authorities, in addition to responsibilities for water supply, sewage treatment, and in some cases navigation, from various other public authorities.

The transfer of responsibilities for pollution control to the ten new regions was undertaken with minimal alteration to the content of the existing water pollution legislation. The main difference was in relation to the consenting of sewage effluent and the potential conflict with the one Authority being both the pollution control authority and the major discharger of effluent. To circumvent the problem of the Authorities issuing consents for their own discharges, provisions were made in the Act for control by the Secretaries of State for the Environment and for Wales.

A.11.   THE CONTROL OF POLLUTION ACT 1974

The Sections of the Control of Pollution Act (COPA) dealing with pollution of water almost entirely repealed the Rivers (Prevention of Pollution) Acts 1951 and 1961, but were not fully implemented until the late 80's. However, the system established by those Acts for water pollution control was on the whole re-enacted in COPA. The more important amendments were the extension of discharge controls to all tidal waters, advertising of applications and public consultation provisions for new discharges, and the creation of a public register of discharge consents and water quality monitoring data. This allowed third parties the opportunity to scrutinise the hitherto confidential information, held by the Regional Water Authorities, regarding consents information, and effluent and water quality data. In cases where the Secretary of State or RWA chose not to prosecute, legal proceedings could be initiated by third parties, using data held on the Register.

A.12.   THE WATER ACT 1989

The main purpose of the 1989 Water Act was to enable the privatisation of the water industry and provided for the division of the Regional Water Authorities into the National Rivers Authority (NRA) and the private Water Service PLC's. The NRA was given the statutory regulatory responsibilities whilst the private water undertakers were allocated the task of provision of water supply and sewage collection and disposal. The separation of the former Regional Water Authorities' regulatory functions, and their operational water supply, sewerage and sewage treatment responsibilities, was achieved with the prime objectives of resolving the conflicting roles of the Water Authorities, and providing the commercial freedom to generate necessary capital investment in the water industry.

Significant additions to pollution control provisions in the Water Act 1989 included provisions for Statutory Water Quality Objectives, Prohibition Notices, Nitrate Sensitive Areas and Water Protection Zones.

A.13   THE ENVIRONMENTAL PROTECTION ACT 1990

The Environmental Protection Act 1990 (EPA) introduced a number of significant new provisions for industrial pollution control. In particular it provided for control of pollution to air water and land by "prescribed" (dangerous) substances discharged from identified "prescribed processes". Implementation of this Act is the responsibility of Her Majesty's Inspectorate of Pollution (HMIP) using the system of "Integrated Pollution Control" (IPC).

Industrial sectors subject to IPC include, amongst others the Fuel and Power, Waste Disposal, Minerals, Chemical and Metal Industries.

The introduction of IPC to these industrial sectors that use "prescribed processes" is being achieved in phases, and HMIP is gradually taking over the responsibilities of the NRA for discharges to controlled waters which arise from prescribed processes. The NRA retains responsibility for monitoring controlled waters affected by discharges from these processes.

A.14.   THE WATER RESOURCES ACT 1991

Part One of the Act and sets out the NRA's functions and general duties. Of particular relevance to the pollution control function is the requirement for the NRA to exercise its powers so as to further the conservation of the natural environment and natural beauty.

Part Three of the Act is concerned with the control of pollution of water resources and is divided into the following relevant sections:

- quality objectives (Sections 82 - 84)

- pollution offences and defences (Sections 85 - 90)

- applications for consents to discharge (Section 88 and Schedule 10)

- appeals (Section 91)

- powers to prevent and control pollution (Sections 92-97)

- supplementary provisions with respect to water pollution (Section 98 -104)

Part Six deals with financial provisions and includes:

- powers to make charges for control of discharges (Sections 131-132)

Part Seven covers land and works powers and includes:

- powers to undertake remedial work, (Sections 161 - 164)

- powers of entry (Sections 169 - 174)

Part Eight specifies information provision requirements and includes:

● public pollution control registers (Section 190)

● provisions relating to the exchange and provision of information (Section 196 -206)

Part Nine covers miscellaneous and supplemental matters and includes:

● requirements for samples to be used as evidence (Section 209)

● byelaw provisions (Section 210 -211)

Appendix B: Summary of EC Directives and International Conventions

There are many EC Directives and International Conventions that have a direct or indirect impact on discharge consents, either through the establishment of quality objectives or emission standards, or through the setting of environmental targets such as reductions in overall emissions. The Sections which follow briefly summarise the key points of those which are most relevant to discharge control.

B.1   EC DIRECTIVES

B.1.1   Introduction

European Community Directives affecting discharges to water include those concerned with setting common standards for environmental quality, emissions, waste treatment and disposal etc. The interpretation and implementation of a Directive is the responsibility of each individual Member State. For a Directive to be effective it must be brought into national legislation, and must be actively applied and enforced. For most Directives the UK regulatory framework already exists under domestic legislation, eg Water Resources Act. The specific requirements of Directives are generally included in Government Regulations issued as Statutory Instruments. These may be supplemented by DoE or MAFF Circulars and Guidance Notes.

The following sections summarise some of the key EC Directives that have impact upon the control of discharges.

B.1.2   EC Directive on Pollution Caused by Certain Dangerous Substances Discharged into the Aquatic Environment of the Community 1976 (76/464/EEC).

Under this framework Directive a series of further (daughter) directives have been adopted addressing specific hazardous substances such as mercury, cadmium, chloroform etc. The requirements of these directives are now enforced in national legislation through the Surface Waters (Dangerous Substances) (Classification) Regulations 1989 and 1992.

These Regulations define environmental quality standards for a range of dangerous substances. They have formally been applied to inland and coastal waters, as statutory Water Quality Objectives, by means of notices served by the Secretary of State in accordance with Section 83 of the Water Resources Act 1991. The NRA is under a duty to use its powers to ensure that such Water Quality Objectives are achieved. The NRA has also been directed by the Secretary of State to ensure that discharges containing substances covered by the Regulations are subject to appropriate consent conditions. The Secretary of State has also directed the NRA as to the monitoring of waters covered by the Regulations.

B.1.3   EC Directive on Urban Waste Water Treatment 1991 (91/271/EEC)

This Directive addresses the collection, treatment and discharge of urban waste water (sewage), and the treatment and discharge of waste waters from certain industrial activities. It also sets out specific requirements for the consenting and monitoring of such effluents. The Directive establishes different levels of sewage treatment necessary (primary, secondary or tertiary) based upon the characteristics or 'sensitivity' of the receiving waters to pollution.

Implementation of this Directive will result in secondary treatment being required for most urban sewage effluents, with primary treatment being acceptable in specified less sensitive coastal and estuarine areas. Crude sewage discharges from populations above a certain size will be eliminated.

The implications of this Directive are covered in more detail in Section 7 of this report.

B.1.4   EC Directive on the Protection of Waters against Pollution Caused by Nitrates from Agricultural Sources 1991 (91/676/EEC)

The Directive requires Member States to identify "polluted waters". For freshwaters, such waters are those where the nitrate limit set has been or could be exceeded. Alternatively the waters may be either eutrophic or have the potential to become eutrophic. Land draining into these waters may need to be designated as vulnerable zones, within which action programmes must be established to reduce and further prevent the agricultural contribution to nitrate pollution. These measures will include rules on the application to land of chemical fertilisers and manure.

B.1.5   EC Directive on the Protection of Groundwater Against Pollution Caused by Certain Dangerous Substances 1980 (80/68/EEC)

This Directive is aimed at eliminating or reducing hazardous substances in groundwater. Due to the inherent problems of monitoring groundwater quality, the emphasis of this Directive is on imposing control measures on the discharge rather than on setting a standard that the receiving water has to achieve. There are no specific monitoring requirements, although visits may be made to determine the effectiveness of the control measures in place.

B.1.6   EC Directive on the Quality of Fresh Waters Needing Protection or Improvement to Support Fish Life 1978 (78/659/EEC)

This Directive was adopted in 1978 and has been implemented through DoE Circulars and guidance notes. It specifies the designation of areas which support fish life, quality parameters and monitoring requirements. The 'Freshwater Fish' Directive provides the baseline quality objectives for pH, ammonia, dissolved oxygen and zinc in a substantial proportion of rivers, approximately 20,000 km of river length in England and Wales. The Directive is a particularly important consideration in determining discharge consents for sewage works. A Water Quality Series report on the Directive is in preparation, for publication in 1994.

B.1.7   EC Directive on the Quality Required of Shellfish Waters 1979 (79/923/EEC)

This Directive was adopted in 1979 and has been implemented through DoE Circulars and guidance notes. It sets quality standards for the protection of shellfish populations, and is aimed at safeguarding the shellfish populations themselves, rather than the health of consumers, (this is covered by the 'Shellfish Health' Directive, 91/492/EEC, below). In England and Wales 18 areas have been designated.

B.1.8   EC Directive Laying Down the Health Conditions for the Production and Placing on the Market of Live Bivalve Molluscs (91/492/EEC)

The 'Shellfish Health' Directive relates to the health of consumers. It specifies bacterial quality standards for classification of shellfish beds and levels of shellfish treatment required prior to marketing. It does not specify water quality criteria. The bacterial quality of shellfish may be affected by sewage discharges, and the locations of outfalls and CSO discharges need to be taken into consideration with respect to classified shellfish beds when sewage improvements are being planned.

B.1.9 Directive on the Quality of Bathing Water 1976 (76/160/EEC)

This Directive addresses the designation of bathing waters and required water quality standards.

The Directive was adopted in December 1975 and Member States were required to improve bathing water standards by December 1985, primarily by the control of sewage discharges. 27 bathing waters were identified in England and Wales in 1980. Following reappraisal, a total of 360 were identified in 1987, and the number currently stands at 419. The requirements of the Directive are now enforced in national legislation through the Bathing Waters (Classification) Regulations 1991.

These Regulations define environmental quality standards for identified bathing waters. They have formally been applied to identified estuarine and coastal waters, as statutory Water Quality Objectives, by means of notice served on the NRA by the Secretary of State in accordance with Section 83 of the Water Resources Act 1991. The NRA is under a duty to use its powers to ensure that the Water Quality Objectives are achieved in each of these bathing waters - primarily through controls placed on sewage discharges. The notice also requires the NRA to monitor compliance with the Regulation requirements.

B.1.10 Directive on Quality Requirements for Surface Waters Intended for the Abstraction of Drinking Water 1975 (75/440/EEC)

This Directive defines different categories of surface water for abstraction of drinking water and associated water quality requirements. These requirements necessarily impact upon discharges upstream of the point of abstraction and the conditions imposed in the consent, to limit the concentration or load of key determinands discharged in the effluent, take account of the assimilative capacity of the waters.

B.2 INTERNATIONAL CONVENTIONS AND AGREEMENTS

B.2.1 Paris Commission

The Paris Commission is responsible for administering the implementation of the recommendations of the Convention for the Prevention of Marine Pollution from Land-based Sources (The Paris Convention, 1974). The convention addressed the persistence, toxicity and bioaccumulation of significant pollutants, such as oils, and outlined means by which this type of pollution could be reduced or ameliorated.

A key output from the convention was a commitment by signatories to maintain or reduce the loads of dangerous substances entering the marine environment. In the UK this has been implemented through the discharge consent mechanism through the issue and/or review of consents, controlling discharges both to freshwater and direct to the marine environment.

The Paris Convention was ratified by the European Community in 1975 and extended to include airborne pollution in 1987.

In 1988 it was agreed that each country involved in the Convention should complete an annual survey of certain substances, including PCBs, specified metals and nutrients, that enter the sea through direct and riverine discharges.

The NRA has the responsibility in England and Wales for undertaking this survey, which was undertaken for the first time in 1990. The survey includes all significant direct industrial and sewage discharges and main river systems in England and Wales, with a designated sampling point just above the tidal limit or confluence. [These latter points will typically be 'Harmonised Monitoring' sites.]

Harmonised Monitoring of river water quality was established in 1974 by the Regional Water Authorities as part of a DoE initiative, and constitutes a national network of water quality sampling sites and a monitoring programme that provides a consistent means for assessing long-term trends in river quality, and an integrated measurement of the input to sea of potential pollutants arising from direct discharges and diffuse inputs within river catchments.

In 1992 the Paris Commission and the Oslo Commission (which controls the dumping at sea of similar substances) were combined into a new Commission to protect the Quality of the North East Atlantic. All the original signatories of the earlier conventions, together with Switzerland, have signed up to the new Convention and ratification is in progress.

B.2.2   Third North Sea Conference

The final declaration of this Ministerial Conference was made in 1990 and contained measures for the protection of the North Sea from land based discharges. A number of the actions to be pursued by the countries bordering the North Sea relate to inputs of dangerous substances that are particularly hazardous due to their persistence, toxicity or bioaccumulation in the environment. It was agreed that discharges of these substances should be reduced to levels that did not pose a risk to either man or the environment by the year 2000. In the interim, reductions of 50% should be achieved for substances input via the rivers and estuaries by the year 1995, as measured against a 1985 baseline.

The NRA is undertaking the appropriate monitoring, of surface water and major effluent discharges, in order to allow the Government to determine the action required and to report progress in reducing inputs of these substances into the North Sea.

APPENDIX C: EXAMPLES OF CONSENTS

EXAMPLE 1: TYPICAL DESCRIPTIVE CONSENT FOR SEWAGE WORKS

NATIONAL RIVERS AUTHORITY   Reference

WATER RESOURCES ACT 1991 - CONSENT TO DISCHARGE

The National Rivers Authority, in pursuance of its powers under the above mentioned Act, HEREBY GIVES CONSENT to the discharge described hereunder subject to the terms and conditions set out below.

Name & Address of Applicant: Any Water Services Co Ltd
123 High Street
Anytown

Date of Application: 18 June 1992

Date of Consent: 10 October 1992

Description of Discharge: Type: Final Effluent

From: Smallish STW

To: Anytown River

Conditions

1 Except with the agreement of the person making the discharge under this consent, no notice shall be served revoking the consent or modifying the conditions before 10 December 1994.

2 The discharge shall consist of treated sewage effluent from an outlet at National Grid Reference XX 8888 8888.

3 The effluent shall derive from domestic sewage from a population of 250 or less and contain no unauthorised trade waste.

4 As far as is reasonably practicable, the works shall be operated so as to prevent:

 a any matter being present in the effluent which will cause the receiving water to be poisonous or injurious to fish or to their spawn, or spawning grounds or food, or otherwise cause damage to the ecology of the receiving waters; and

 b the treated effluent from having any other adverse environmental impact.

5 The Company will operate the works having regard, so far as is relevant, to the guidance set out in the National Water Council's Occasional Technical Paper Number 4, "The Operation and Maintenance of Small Sewage Treatment Works" dated January 1980. In particular, the works shall be maintained properly such that:

 a it remains fully operational except at time of mechanical or electrical breakdown;

 b any such breakdowns shall be attended to promptly and the equipment returned to normal operation as soon as possible; and

 c tanks shall be regularly desludged at sufficient frequency and in such a manner as to prevent problems with septic tanks, rising sludge or excessive carryover of suspended solids.

6 Facilities shall be provided for safe and convenient access to enable Authority's representatives at any time to take samples, carry out flow measurements and inspection to ensure that the conditions of this consent are complied with.

NRA Regional Office

Address   NRA Authorised Signatory

NATIONAL RIVERS AUTHORITY   **Reference**

WATER RESOURCES ACT 1991 - CONSENT TO DISCHARGE

The National Rivers Authority, in pursuance of its powers under the above mentioned Act, HEREBY GIVES CONSENT to the discharge described hereunder subject to the terms and conditions set out below.

Name & Address of Applicant: Any Water Services Co Ltd
                                      123 High Street
                                      Anytown

Date of Application: 18 July 1992

Date of Consent: 10 November 1992

Description of Discharge: Type: Final Sewage Effluent

                                      From: Somewhere STW

                                      To: Somewhere Stream

This consent shall not be taken as providing a statutory defence against a charge of pollution in respect of any poisonous, noxious or polluting constituents not specified herein.

Conditions

1   General

    a   This consent shall come into force on 1 May 1993.

    b   Except with the agreement of the person making the discharge under this consent, no notice shall be served revoking the consent or modifying the conditions before 1 May 1995.

    c   For the purpose of applying the conditions identified in section 3 below, the discharger shall provide and maintain facilities which will enable the Authority's representatives to take flow measurements of the final sewage effluent which is discharged at the outlet.

        The discharger shall identify the facility with a clearly visible sign, distinguishing it from any other and provide a clearly visible notch, mark, or device indicating the level equivalent to the maximum instantaneous consented flow.

    d   For the purpose of applying the conditions identified in section 4 below, the discharger shall provide and maintain facilities which will enable the Authority's representatives to take discrete samples of the final sewage effluent which is discharged at the outlet. The discharger shall identify the facility with a clearly visible sign distinguishing it from any other.

    e   The discharger shall provide to the Authority's satisfaction a drawing showing the precise location of the facilities provided in accordance with conditions (c) and (d) above not later than one month prior to the date of enforcement of this consent.

    f   Facilities shall be provided for safe and convenient access to enable the Authority's representatives at any time to take samples, carry out flow measurements and inspection to ensure that the conditions of this consent are complied with.

2   As to Outlet

    An outfall shall be sited at NGR XX 9999 9999 and shall be so constructed that it is used for the discharge of final sewage effluent derived only from this sewage treatment works.

3   As to Discharge

a   The maximum instantaneous rate of discharge shall not exceed 1.18 litres per second.

    b   The volume discharged under dry weather flow conditions shall not exceed 39.6 cubic metres in any period of twenty four hours.

4   As to Discharge Composition

   a   The discharge shall:

      i   contain no visible signs of oil or grease.

      ii   at no time contain any matter, other than matter specifically authorised or limited by numerical conditions in this consent, to such an extent as to cause the receiving waters to be poisonous or injurious to fish or the spawning ground, spawn or food of fish.

   b   In any series of samples of the final effluent taken over any twelve month period as listed in column 1 of the table set out in the annex to this schedule, then, in respect of the following determinands, no more than the relevant number as permitted in column 2 of the table shall be:

      i   in excess of 40 milligrams per litre of biochemical oxygen demand (BOD) measured after 5 days at 20°C with nitrification suppressed by the addition of allyl thiourea;

      ii   in excess of 60 milligrams per litre of suspended solids (measured after drying for one hour at 105°C);

      iii   in excess of 10 milligrams per litre of ammoniacal nitrogen expressed as nitrogen.

   c   No single sample of the final effluent discharged shall have:

      i   in excess of 80 milligrams per litre of biochemical oxygen demand (BOD) measured after 5 days at 20°C with nitrification suppressed by the addition of allyl thiourea;

      ii   in excess of 120 milligrams per litre of suspended solids (measured after drying for one hour at 105°C);

      iii   in excess of 20 milligrams per litre of ammoniacal nitrogen expressed as nitrogen;

      iv   a pH value less than 6 or greater than 9.

NRA Regional Office

Address   NRA Authorised Signatory
                                                         NRA Region

ANNEX 1: TABLE

Column 1 Series of samples taken in any period of 12 months	Column 2 Maximum number of samples for given determinand permitted to exceed limit
4- 7	1
8- 16	2
17- 28	3
29- 40	4
41- 53	5
54- 67	6
68- 81	7
82- 95	8
96-110	9
111-125	10
126-140	11
141-155	12
156-171	13
172-187	14
188-203	15
204-219	16
220-235	17
236-251	18
252-268	19
269-284	20
285-300	21
301-317	22
318-334	23
335-350	24
351-365	25